Maik Dittmann

Isogeometric analysis and hierarchical refinement for multi-field contact problems

**Schriftenreihe des Instituts für Mechanik
Karlsruher Institut für Technologie (KIT)**

Band 4

Herausgeber:
Prof. Dr.-Ing. habil. Peter Betsch
Prof. Dr.-Ing. habil. Thomas Seelig

Eine Übersicht aller bisher in dieser Schriftenreihe erschienenen Bände finden Sie am Ende des Buches.

Isogeometric analysis and hierarchical refinement for multi-field contact problems

by
Maik Dittmann

KIT Scientific Publishing

Dissertation, Universität Siegen
Naturwissenschaftlich-Technische Fakultät
Tag der mündlichen Prüfung: 09. November 2016
Hauptreferentin: Prof. Dr.-Ing. Kerstin Weinberg
Korreferenten: PD Dr.-Ing. Christian Hesch, Prof. Dr.-Ing. Udo Nackenhorst

Impressum

KIT Scientific Publishing

Karlsruher Institut für Technologie (KIT)
KIT Scientific Publishing
Straße am Forum 2
D-76131 Karlsruhe

KIT Scientific Publishing is a registered trademark of Karlsruhe Institute of Technology. Reprint using the book cover is not allowed.

www.ksp.kit.edu

Print on Demand 2017 – Gedruckt auf FSC-zertifiziertem Papier

ISSN 2363-4936
ISBN 978-3-7315-0616-4
DOI 10.5445/KSP/1000063914

Isogeometric analysis and hierarchical refinement for multi-field contact problems

DISSERTATION

zur Erlangung des Grades eines Doktors

der Ingenieurwissenschaften

vorgelegt von

M.Sc. Maik Dittmann

geb. am 19.01.1984 in Kirchen

eingereicht bei der Naturwissenschaftlich-Technischen Fakultät

der Universität Siegen

Siegen 2016

Hauptreferentin:	Prof. Dr.-Ing. Kerstin Weinberg
Korreferenten:	PD Dr.-Ing. Christian Hesch
	Prof. Dr.-Ing. Udo Nackenhorst
Tag der mündl. Prüfung:	09.11.2016

Abstract

The present work deals with thermomechanical and fracture mechanical contact problems in the context of isogeometric analysis. In particular, a novel formulation of the thermomechanical system based on a Hellinger-Reissner type variational principle is introduced, whereas a phase-field approach is focused for the fracture mechanical system using a newly defined anisotropic split of the principal invariants. To improve accuracy and convergence, a phase-field model of fourth order is considered requiring an at least C^1-continues discretization throughout the domain. An isogeometric framework is addressed for the spatial discretization subject to hierarchical refinements to resolve local features. The corresponding discrete contact formulations are based on a variationally consistent mortar contact approach adapted for non-uniform rational B-spline discretized and hierarchical refined contact surfaces. For a proper energy transfer between the mechanical and the thermal field, newly defined triple mortar integrals are introduced as well. The capabilities of the proposed framework are demonstrated within numerous numerical examples.

Keywords: Isogeometric analysis, thermomechanics, fracture mechanics, phase-field approach, mortar contact

Kurzfassung

Die vorliegende Arbeit befasst sich mit thermomechanischen und bruchmechanischen Kontaktproblemen im Rahmen der isogeometrischen Analyse. Dabei wird eine neuartige, auf dem Hellinger-Reissner Variationsprinzip basierende Formulierung für das thermomechanische System eingeführt, während für das bruchmechanische System ein Phasenfeld Ansatz zusammen mit einer neu definierten anisotropen Zerlegung der Invarianten verwendet wird. Zur Verbesserung der Genauigkeit und der Konvergenz, wird ein Modell vierter Ordnung für das Phasenfeld herangezogen, welches eine global mindestens C^1-stetige räumliche Diskretisierung erfordert. Unter anderem wird daher auf das isogeometrische Konzept zurückgegriffen, welches durch einen hierarchischen Ansatz zur lokalen Verfeinerung ergänzt wird. Die entsprechenden diskreten Kontaktformulierungen basieren auf einem variationell konsistenten Mortar Kontaktansatz, angepasst für NURBS diskretisierte und hierarchisch verfeinerte Kontaktflächen. Des Weiteren werden für einen ordnungsgemäßen Energietransfer zwischen dem mechanischen und dem thermischen Feld neu definierte dreifach Mortar Integrale eingeführt. Die Leistungsfähigkeit des vorgeschlagenen Gesamtansatzes wird durch eine Vielzahl von numerischen Beispielen demonstriert.

Schlüsselwörter: Isogeometrische Analyse, Thermomechanik, Bruchmechanik, Phasenfeld Ansatz, Mortar Kontakt

Preface

The work presented in this thesis was carried out during my time as research assistant at the Chair of Solid Mechanics at the University of Siegen from 2012 to 2016. Financial support for this research was provided by the Deutsche Forschungsgemeinschaft (DFG) under grants HE 5943/3-1 and HE 5943/6-1 which I gratefully acknowledge.

First of all I want to thank my advisor Doctor Christian Hesch for providing me the doctoral position and the research topic. His outstanding support as well as his motivation throughout the whole time of my doctoral study allowed for a quick and successful treatment of the research topic. I would like to express my gratitude to Professor Kerstin Weinberg for letting me be a part of her group and for acting as my referee. Moreover, I thank Professor Udo Nackenhorst for his interest in my work and his willingness to serve as co-referee. Also many thanks to Professor Thomas Carolus and the head of the committee Professor Oliver Nelles.

I thank Professor Peter Betsch who gave me the opportunity to start my doctoral study.

Next, I would like to thank all my colleagues inside and outside the chair for an excellent collaboration. Especially, it was a great pleasure to me to work on the ESRA-project together with Doctor Christian Hesch, Doctor Marlon Franke, Stefan Schuß, Carola Bilgen, Jonathan Schulte, Friedemann Streich, Alexander Janz, Marek Werner and Mark Schiebl. I am grateful to Stefan Schuß for fruitful discussions on some mathematical issues. Regarding this thesis, my thanks also go to Doctor Marlon Franke, Carola Bilgen, Marek Werner and Jonathan Schulte for stylistic and linguistic suggestions. In addition, I would like to thank Gisela Thomas and Harald Schmitz for their always friendly guidance and support in many off-topic matters.

Finally, I am most thankful for the continuous support and patience of my beloved wife Svenja and our daughter Jolina throughout the whole time of my academic education and doctoral work.

Siegen, September 2016 Maik Dittmann

Contents

List of Figures

List of Tables

1 Introduction

Over recent decades, the demand for computational simulations and their complexity have been increased. On the other hand, the dramatically rising computational power allows for more and more realistic simulations of complex machine parts or of biological systems, to name only a few applications. In particular, large deformations and displacements along with thermomechanical or electromechanical effects are often considered within this context. Moreover, computational contact methods accounting for the multiphysical character of complex industrial applications are most crucial nowadays for the numerical analysis of entire technical designs. New challenges arise due to actual developments in the field of fracture mechanics, which are in a natural way part of a comprehensive analysis of contact and impact problems and have to be included within a corresponding numerical framework.

For the numerical simulations using a finite element framework, efficient, accurate and robust discretization methods, which fulfill the respective continuity requirements of the underlying mathematical problem are necessary. In a nutshell, isogeometric analysis (IGA) using non-uniform rational B-spline (NURBS) based shape functions is able to account for higher order models and provides an exact representation of computer aided design (CAD) geometries. It is often emphasized that CAD data can be directly transferred into finite element analyzes (FEA) to define the computational mesh. This is not true in general since the respective requirements are often not the same. To be specific, local refinement procedures are necessary to resolve locally sharp gradients or other local areas of interest within large-scale three dimensional systems without violating the given continuity requirements.

1.1 State of the art and objectives

In the present work we aim at an isogeometric framework for thermomechanical as well as fracture mechanical large deformation contact problems[1].

1.1.1 Isogeometric analysis

Assuming the existence of a specific weak form defined in a domain, Galerkin type methods based on shape functions with global support have already been applied to simple

[1] The work summarizes several articles [3, 27, 42, 44, 43] which have been published by the author.

geometries more than half a century ago. Subdividing the considered domain into a finite set of non-overlapping elements as proposed by Courant [21] and later on by Turner et al. [94] allows for the use of shape functions with local support in each element and thus, for a simple approximative representation of complex geometries. Such finite element approximations found numerous applications in various disciplines, although it comes at the cost of reduced continuity across the element boundaries. About a decade ago T. Hughes [49, 20, 19] introduced NURBS as finite element basis functions with the aim to gain control on continuity requirements in design and to maintain the geometrical exact representation of complex engineering structures (see also Cottrell et al. [18] for a comprehensive review of the concept). He coined the term isogeometric analysis which we will continue to use for all NURBS based finite element methods.

B-splines and NURBS have local support but are not restricted to a single finite element. In the multivariate case they have a tensor product structure, which turn into major drawbacks for the construction of local refinement procedures. In order to solve these mesh subdivision problems T-splines have been introduced in the IGA to break the tensor product structure of the spline base, see Bazilevs et al. [9]. Because additional knots are inserted into the structure, the arising T-junctions can be considered as the IGA version of hanging nodes of standard FEA. As shown in Scott et al. [83] such refinement algorithms improve the quality of approximation but become very complex in three dimensional domains. Additionally, restriction on the polynomial degree, on the linear dependence of T-spline blending functions and on the locality of the refinement are known.

An alternative to the element based mesh subdivision procedures provides the hierarchical superposition of shape functions. Such refinement methods are long known from high-order FEA, see, e.g. the work of Szabò & Babuška [89], and for B-splines they have already been introduced by Forsey & Bartels [30]. But only recently such refinements have been applied within the framework of IGA, see, e.g. Vuong et al. [95] and Evans et al. [29]. The basic idea behind a hierarchical refinement is simple: B-spline and NURBS basis functions are replaced on the refined level by a linear combination of scaled and copied versions of themselves, i.e. the refinement concept relies on a subdivision of the underlying B-spline and NURBS functions. This procedure maintains the global properties in terms of smoothness and continuity of the unrefined mesh at the cost of a non-trivial implementation. Extensions to the multivariate case are possible. Although the general concept of hierarchical refinement is straightforward, in detail the realizations within the FEA differ.

Here, we deal with a novel hierarchical scheme which provides an efficient refinement for arbitrary one, two and three dimensional B-spline and NURBS bases and additionally preserves the partition of unity property of the finite element basis functions and the smooth transitions to the original mesh for every degree of continuity, see Hesch et al. [44].

1.1.2 Large strain thermoelasticity

Thermoelastic material models have been investigated over the past decades, see Reese and Govindjee [79] and Miehe [63] among many others. A comprehensive survey can be found in textbooks like Malvern [60]. In Holzapfel and Simo [47] the research on rubber elasticity has been extended to a class of entropic elastic materials, written entirely in the material configuration.

Based on the original ideas proposed by Schröder et al. [81], a novel formulation for large strain polyconvex elasticity was recently introduced in Bonet et al. [10]. Within the formulation conjugate stresses to an extended kinematic set, i.e. the deformation gradient, its co-factor and its Jacobian determinant, are introduced to define a generalized convex complementary energy function and a corresponding complementary energy principle of the Hellinger–Reissner type, where the conjugate stresses along with the deformed geometry are primal variables. See also Washizu [97] for a Hu-Washizu type of mixed variational principle.

Within the present work we extend the formulation introduced in Bonet et al. [10] to large strain thermoelasticity. Therein, the set of primal variables is supplemented by the absolute temperature of the system. The thermal field itself is directly coupled with the conjugate stress to the Jacobian determinant, whereas the dependence to the deformation is only given via the definition of the heat conduction. In the context of thermoelasticity such a formulation has not been addressed so far, at least to the best knowledge of the author. For the numerical simulations, the deformed geometry and the temperature are discretized using the isogeometric concept, whereas discontinuous linear and constant interpolations of the stresses allow for a condensation procedure to reduce the global system.

1.1.3 Phase field approach to fracture

The numerical prediction of fracture patterns play an important role in engineering design. To determine the actual fracture pattern, Griffith and Irwin [34, 51] formulated a most general concept for brittle fracture models by assuming that the material fails locally upon the attainment of a specific fracture energy regarding a critical energy release rate. Due to the complexity of the evolving fracture surfaces in three dimensional solids, the evaluation of appropriate interface conditions is a most challenging task within a finite element framework.

Since a straightforward minimization of the fracture energy is not feasible, more generalized numerical formulations have been introduced to avoid the algorithmic tracking of discontinuities within the solid. In particular, phase-field methods provide a most general framework to formulate variational fracture models based on energy functionals embedded within a global minimizer. In contrast to classical discontinuous methods, such as interface element formulations [16, 71, 72] or local enrichment strategies [86, 35], phase-field methods as formulated by Francfort and Marigo [31] introduce a non-local diffusive crack zone on the finite element mesh. Embedded within a continuum mechanical framework as proposed by Miehe et al. [66], the order parameter of the phase-field allows to characterizes the current state of fracture within an arbitrary shaped three dimensional solid.

For large strain problems, the hyperelastic material model is degenerated by the phase-field. Independent of the current state of fracture, the model has to satisfy certain convexity criteria in order to guarantee the existence of minimizers, see [61]. One of the most important and well-established criteria is the polyconvexity criteria in the sense of Ball [8]. This concept has been adapted to a large variety of applications, see [36, 81, 70].

We introduce a new formulation based on an anisotropic split of the principle invariants, which satisfies the polyconvexity criteria for the mechanical field independent of the current state of fracture. This is of major importance, since we can avoid the application of trust region and similar optimization tools, see Hesch et al. [43]. Eventually, we apply a fourth-order model of the phase-field as well, which requires an at least C^1-continuous spatial discretization, see also Borden et al. [12] and Weinberg and Hesch [98]. Combined with the introduced formulation on principle invariants for finite strains, this most general methodology is expected to improve the accuracy and the convergence of the numerical solution in finite elasticity.

1.1.4 Contact mechanics

From an engineering point of view contact mechanics has a long history, but just in recent decades the non-linear contact constraints was directly incorporated within numerical simulations. Several contact formulations have been investigated, see, e.g. Nackenhorst [68] and Suwannachit [88] for rolling contact problems using an arbitrary Lagrangian Eulerian formulation and Franke et al. [32] for large deformation frictional contact problems utilizing an augmentation technique. Formulations for thermomechanical frictional contact and impact problems have been investigated and analyzed in, e.g. Zavarise et al. [103], Strömberg et al. [87], Saracibar [1] and Laursen [56]. Contact problems for IGA have been addressed in a series of papers throughout the past years, see De Lorenzis et al. [25, 26] and Temizer et al. [92, 90, 93]. In these works, a knot-to-surface (KTS) method has been developed and extended to mortar based contact formulations, see also Temizer and Hesch [91] for frictionless mortar based contact formulations in the context of hierarchical refinements. Matzen et al. [62] have proposed a collocation based approach, analogous to the well-known node-to-surface (NTS) method. See also Kim and Youn [54] for a mortar approach and Lu [58] for an alternative contact treatment.

Mortar formulations in the context of IGA domain decomposition problems have been presented in Hesch and Betsch [41]. In the present work, we extend the ideas developed therein to thermomechanical contact in order to achieve a variationally consistent mortar formulation for the discrete, hierarchical refined contact interface. In particular, the mortar projections will be calculated via a newly developed segmentation procedure of the surface intersections, see Puso et al. [78] for a discussion about different spatial integration schemes. For the thermal contributions, triple mortar integrals will be introduced to accurately capture the frictional dissipation contribution to the contact heat flux and to establish a correct thermal interaction among the contacting surfaces, see Dittmann et al. [27]. See also Hüeber and Wohlmuth [48] for mortar methods applied to thermoelasticity.

The mortar contact formulation for thermomechanical problems can be adapted to phase-field fracture problems in a straightforward manner, since the contact surface does not interfere with the phase-field boundary. The application of contact algorithms to phase-field fracture problems has been addressed for the first time in Hesch et al. [42], at least to the knowledge of the authors.

1.2 Organization of the work

The outline of the work is as follows. The concept of IGA along with a hierarchical refinement procedure based on spline subdivision is outlined in chapter 2. Large strain thermoelasticity including a Hellinger-Reissner type variational principle is addressed in chapter 3, whereas in chapter 4 a large strain phase-field approach to fracture mechanics is presented. In chapter 5 we deal with thermomechanical as well as fracture mechanical contact problems. Representative examples are provided at the end of each of these chapters. Finally, conclusions are drawn in chapter 6.

2 Isogeometric concept

In this chapter we provide an introduction of the isogeometric concept for general FEA. In particular, we employ NURBS (see Piegl and Tiller [75] for an efficient construction) as basis functions for the spatial discretization within a finite element framework (see also Cottrell et al. [18]). This allows us to control the continuity of the finite element approximation and to provide an exact geometrical representation based on CAD data. Moreover, a hierarchical approach based on spline subdivision is introduced to enable local refinements which are vital to resolve local features within large scale three dimensional simulations, see, e.g. Hesch et al. [42, 44]. Local refinement procedures in the context of IGA are subject to current research projects (see, e.g. [9, 96, 80, 83, 13]), where the present approach maintains the continuity of the unrefined spline bases as well as the partition of unity and can be applied to arbitrary B-spline and NURBS based approximations. Eventually, a transient Kuramoto-Sivashinsky problem (see [73, 100, 33, 3]) is considered to demonstrate the capabilities of the hierarchical refinement scheme. This transient partial differential equation of fourth order allows us to investigate the consistency as well as the convergence of the proposed refinement scheme, cf. Hesch et al. [44].

2.1 Basics

Starting with the univariate case in this section we demonstrate the recursive construction of B-spline basis functions and outline the most important properties of B-spline bases for FEA, see also Hughes et al. [49], among many others. The extension to multivariate B-spline bases relies on the tensor product of univariate bases, and therefore all properties are also valid in the multivariate case. Moreover, we introduce NURBS as rational functions of B-splines and deduce the corresponding derivatives.

2.1.1 Univariate B-splines

We consider a univariate B-spline basis of polynomial degree p defined by the knot vector $\Xi = [\xi_1, \xi_2, \ldots, \xi_{\mathfrak{n}+p+1}]$, where $\mathfrak{n}$ is the number of B-spline basis functions. Assuming that Ξ is a non-decreasing set of coordinates in the parameter space, i.e. $\xi_i \leq \xi_{i+1}$, the recursive definition of a single univariate B-spline is given as follows[1]

$$B_p^i(\xi) = \frac{\xi - \xi_i}{\xi_{i+p} - \xi_i} B_{p-1}^i(\xi) + \frac{\xi_{i+p+1} - \xi}{\xi_{i+p+1} - \xi_{i+1}} B_{p-1}^{i+1}(\xi), \tag{2.1}$$

[1] In case of repeated knots we need to define $\frac{0}{0} := 0$ for the recursion.

beginning with

$$B_0^i(\xi) = \begin{cases} 1 & \text{if } \xi_i \leq \xi < \xi_{i+1} \\ 0 & \text{otherwise} \end{cases} . \tag{2.2}$$

Each B-spline B_p^i is a piecewise positive polynomial with support on $[\xi_i, \xi_{i+p+1})$. In addition, we introduce a modified knot vector $\tilde{\Xi} = [\tilde{\xi}_1, \ldots, \tilde{\xi}_{\tilde{\mathfrak{n}}+1}]$ without any repetitions and specify the knot multiplicity m_j related to $\tilde{\xi}_j$ as its corresponding number of repetitions in Ξ, cf. Brivadis et al. [14]. Now, the partition $\tilde{\mathcal{B}}_j = [\tilde{\xi}_j, \tilde{\xi}_{j+1})$ yields the jth element of the mesh in the parametric interval $\tilde{\mathcal{B}} = [\tilde{\xi}_1, \tilde{\xi}_{\tilde{\mathfrak{n}}+1})$, where $\tilde{\mathfrak{n}}$ denotes the number of elements.

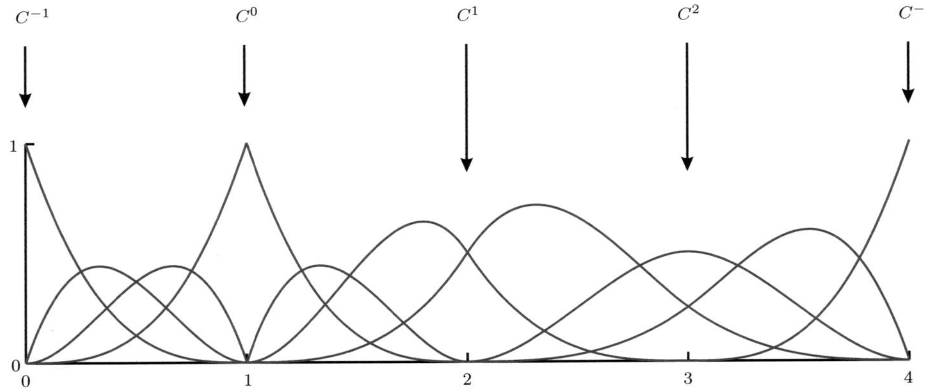

Figure 2.1: Continuity of a cubic B-spline basis ($p = 3$) defined by the knot vector $\Xi = [0, 0, 0, 0, 1, 1, 1, 2, 2, 3, 4, 4, 4, 4]$ and $\tilde{\Xi} = [0, 1, 2, 3, 4]$, respectively.

As shown in Figure 2.1, B-splines are C^∞-continuous within the elements, whereas C^{p-m_j}-continuity[2] occurs at $\tilde{\xi}_j$. Note that in general a B-spline basis is not interpolatory, only if the inter element continuity is less than C^1, i.e. the knot multiplicity is higher than $p-1$, the basis becomes interpolatory at the corresponding knot. Concerning FEA a B-spline basis satisfies two additional important properties, which are:

- Partition of unity $\sum_i B^i(\xi) \equiv 1$
- Linear independence $\sum_i B^i(\xi) c_i \equiv 0 \quad \Leftrightarrow \quad c_i = 0$

Please note that the former requirement can be relaxed at the cost of introducing a subsequent normalization scheme (see, e.g. Temizer & Hesch [91]), whereas the latter one is proved in De Boor [24]. The knot vector used in Figure 2.1 is referred to be open, since the knot multiplicity related to the first and last knot in $\tilde{\Xi}$ is equal to $p+1$. For the application of a non-open knot vector, e.g. a uniform knot vector, we need to consider that $\tilde{\xi}_1 < \xi_{p+1}$ and $\tilde{\xi}_{\tilde{\mathfrak{n}}+1} > \xi_{\mathfrak{n}+1}$, respectively, since a set of B-splines constitutes a valid

[2] C^m-continuity of the approximation spaces relates to solution functions u which are at least $u \in \mathcal{H}^{m+1}(\mathcal{B})$, where $\mathcal{H}^{m+1}(\mathcal{B})$ is the Sobolev space of square integrable functions with $m+1$ square integrable derivatives.

basis only on $\tilde{\mathcal{B}}_\text{b} = [\xi_{p+1}, \xi_{\text{n}+1}$. This is evident in Figure 2.2, where the full set of $p+1$ B-splines which have non-zero values is not available outside the interval $[\xi_{p+1}, \xi_{\text{n}+1})$, and therefore the partition of unity is not fulfilled.

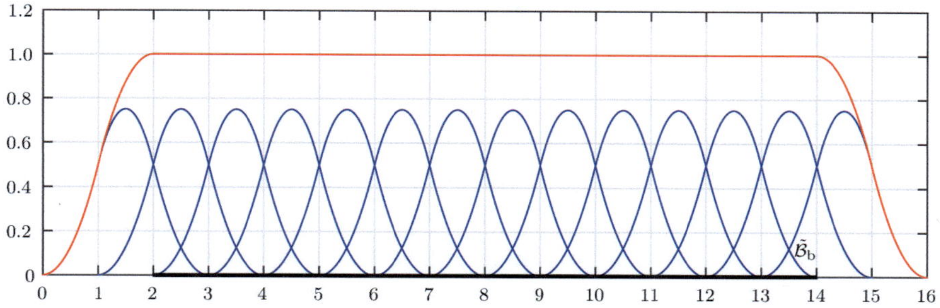

Figure 2.2: Quadratic B-spline basis ($p = 2$) defined by a uniform knot vector $\Xi = \tilde{\Xi}$. The red line represents the sum of B-spline basis functions.

Note that the outlined properties of univariate B-spline basis functions are also valid for multivariate B-spline basis functions as well as for NURBS basis functions.

Derivatives

The derivatives of B-splines may be expressed in an efficient way by utilizing B-splines of lower order. We obtain the first derivative of the ith B-spline basis function with

$$\frac{\mathrm{d}}{\mathrm{d}\xi} B_p^i(\xi) = \frac{p}{\xi_{i+p} - \xi_i} B_{p-1}^i(\xi) - \frac{p}{\xi_{i+p+1} - \xi_{i+1}} B_{p-1}^{i+1}(\xi) \tag{2.3}$$

and consequently the kth derivative with

$$\frac{\mathrm{d}^k}{\mathrm{d}\xi^k} B_p^i(\xi) = \frac{p}{\xi_{i+p} - \xi_i} \left(\frac{\mathrm{d}^{k-1}}{\mathrm{d}\xi^{k-1}} B_{p-1}^i(\xi) \right) - \frac{p}{\xi_{i+p+1} - \xi_{i+1}} \left(\frac{\mathrm{d}^{k-1}}{\mathrm{d}\xi^{k-1}} B_{p-1}^{i+1}(\xi) \right), \tag{2.4}$$

see Cottrell et al. [18] for an expression purely in terms of lower order functions.

2.1.2 Multivariate B-splines

With the definition of a univariate B-spline basis at hand we can create a multivariate B-spline basis of degree $\boldsymbol{p} = [p_1, \ldots, p_d]$ and dimension $d \in \{2, 3\}$. To be specific, the multivariate basis is defined by the dyadic product $\boldsymbol{\Xi} = \Xi_1 \otimes \ldots \otimes \Xi_d$ of univariate knot vectors Ξ_l, where $l \in \{1, \ldots, d\}$ denotes the direction in the parameter space. Next, we apply the dyadic product to the modified knot vectors $\tilde{\Xi}_l$, so that the partition $\tilde{\mathcal{B}}_j = [\tilde{\xi}^1_{j_1}, \tilde{\xi}^1_{j_1+1}) \times \ldots \times [\tilde{\xi}^d_{j_d}, \tilde{\xi}^d_{j_d+1})$ forms an element of the mesh in the parametric domain $\tilde{\mathcal{B}}$.

Eventually, a multivariate B-spline is given by[3]

$$B^A = B_{\boldsymbol{p}}^{\boldsymbol{i}}(\boldsymbol{\xi}) = \prod_{l=1}^{d} B_{p_l}^{i_l}(\xi^l), \tag{2.5}$$

with the multi index $\boldsymbol{i} = [i_1, \ldots, i_d]$ and $\text{supp}(B^A) = [\xi^1_{i_1}, \xi^1_{i_1+p_1+1}) \times \ldots \times [\xi^d_{i_d}, \xi^d_{i_d+p_d+1})$, see Figure 2.3 for illustration.

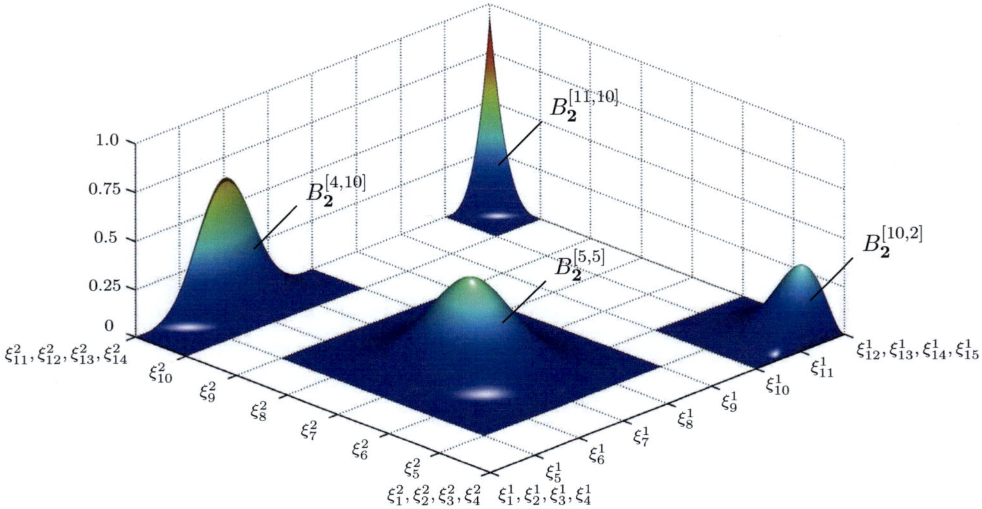

Figure 2.3: Illustration of the multi index and the support of single cubic B-spline basis functions in 2D. The mesh in the parameter space is given by the dyadic product of modified knot vectors $\tilde{\boldsymbol{\Xi}} = [\tilde{\xi}^1_1, \ldots, \tilde{\xi}^1_9] \otimes [\tilde{\xi}^2_1, \ldots, \tilde{\xi}^2_8]$.

Derivatives

The partial derivative of the Ath multivariate B-spline with respect to ξ^l reads

$$\frac{\partial B^A}{\partial \xi^l} = \frac{\partial}{\partial \xi^l} B_{\boldsymbol{p}}^{\boldsymbol{i}}(\boldsymbol{\xi}) = \frac{\partial}{\partial \xi^l} B_{p_l}^{i_l}(\xi^l) \prod_{\substack{\hat{l}=1 \\ \hat{l}\neq l}}^{d} B_{p_{\hat{l}}}^{i_{\hat{l}}}(\xi^{\hat{l}}). \tag{2.6}$$

Higher order derivatives can be obtained in a straightforward manner by using (2.4) instead of (2.3).

[3] The index A denotes the global basis function number. Its assignment to the multi index $\boldsymbol{i}$ can be found in Cottrell et al. [18].

2.1.3 NURBS

Multivariate NURBS are constructed as rational functions of multivariate B-spline functions[4]

$$R^A = R_{\boldsymbol{p}}^{\boldsymbol{i}}(\boldsymbol{\xi}) = \frac{\prod_{l=1}^{d} B_{p_l}^{i_l}(\xi^l) w_{\boldsymbol{i}}}{\sum_{\boldsymbol{j}} \prod_{l=1}^{d} B_{p_l}^{j_l}(\xi^l) w_{\boldsymbol{j}}}, \tag{2.7}$$

where the denominator is referred to as the weight function and $w_{\boldsymbol{i}}$ denotes positive NURBS weights. Note that if each weight has the same value, i.e. $w_{\boldsymbol{i}} / \sum_{\boldsymbol{j}} B_{\boldsymbol{p}}^{\boldsymbol{j}}(\boldsymbol{\xi}) w_{\boldsymbol{j}} = 1 \,\forall\, \boldsymbol{\xi}$, we will obtain again the original B-spline. Thus, B-splines can be considered as a special case of NURBS.

Derivatives

The partial derivative of R^A with respect to ξ^l can be obtained in a straightforward manner by applying the quotient rule

$$\frac{\partial R^A}{\partial \xi^l} = \frac{\sum_B B^B w_B \frac{\partial B^A}{\partial \xi^l} - B^A \sum_B \frac{\partial B^B}{\partial \xi^l} w_B}{\left(\sum_B B^B w_B\right)^2} w_A. \tag{2.8}$$

Moreover, the corresponding derivative of order k is given by

$$\frac{\partial^k R^A}{\partial {\xi^l}^k} = \frac{\frac{\partial^k B^A}{\partial {\xi^l}^k} w_A - \sum_{m=1}^{k} \binom{k}{m} \frac{\partial^{k-m} R^A}{\partial {\xi^l}^{k-m}} \sum_B \frac{\partial^m B^B}{\partial {\xi^l}^m} w_B}{\sum_B B^B w_B}. \tag{2.9}$$

2.2 Hierarchical refinement of B-splines

In the following we provide a proposal to an intuitive, efficient and most general construction of hierarchical refinements. Moreover, an illustrative example clarifies the novel hierarchical scheme and emphasizes important properties as linear independence and partition of unity. The extension to non-uniform B-splines as well as NURBS can be found in section 2.4.

2.2.1 Subdivision of uniform B-splines

A single B-spline B^A defined by equidistantly spaced knot sequences $[\xi_{i_l}^l, \ldots, \xi_{i_l+p_l+1}^l]$ and their dyadic product, respectively, can be represented as a linear combination of scaled,

[4] For the definition of NURBS and their derivatives we do not use the Einstein notation.

contracted and shifted copies of itself as

$$B_{\boldsymbol{p}}^{\boldsymbol{i}}(\boldsymbol{\xi}) = \sum_{\boldsymbol{j}=\boldsymbol{0}}^{\boldsymbol{p}+\boldsymbol{1}} \prod_{l=1}^{d} 2^{-p_l} \binom{p_l+1}{j_l} B_{p_l}^{i_l}(2\xi^l - j_l h_l - \xi_{i_l}^l), \tag{2.10}$$

where $\mathbf{1} = [1, \dots, 1] \in \mathbb{R}^d$ and h_l is the uniform knot span width in the lth parametric direction. The original B-spline is then referred to as the parent of its $(p_1+2)\cdot\ldots\cdot(p_d+2)$ children. Unless otherwise specified, we do not consider the univariate case separately in the following, hence we set $d \in \{1, 2, 3\}$. Now, the subdivision procedure related to (2.10) is illustrated for univariate B-splines of different order in Figure 2.4, whereas in Figure 2.5 the corresponding procedure for the two dimensional case is shown. As we can see, the approach subdivides the parametric domain into half step size to provide sufficient positions for the construction of B-splines referred to as children.

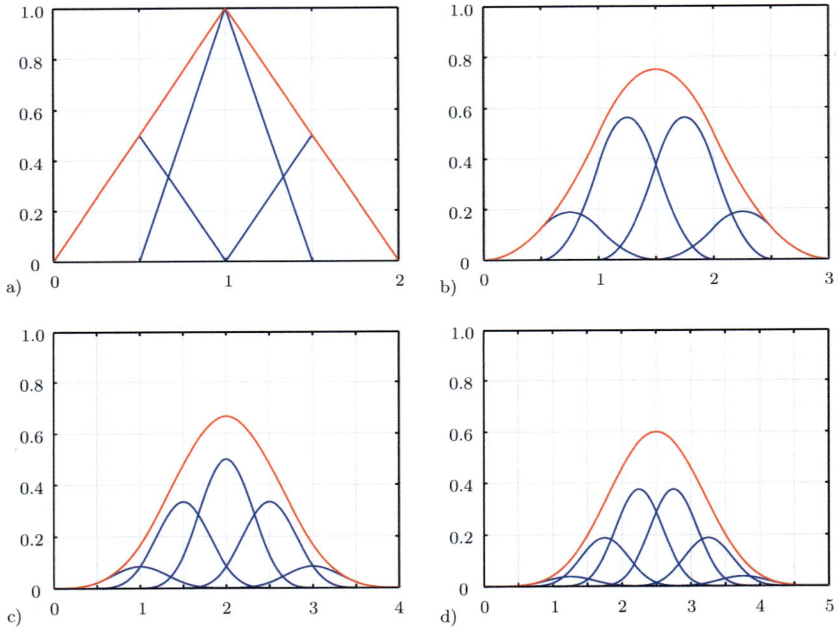

Figure 2.4: Subdivision of a single a) linear, b) quadratic, c) cubic and d) quartic B-spline in 1D. The parent B-spline (red line) corresponds to the sum of its scaled children (blue line).

2.2.2 Hierarchical basis

Our starting point is a uniform set of B-splines to be refined locally, see, e.g. Figure 2.2. The associated functions are then referred to as level 0 B-splines $B^{0,A}$ listed in the index set $\mathcal{I}^0$. Furthermore, the support of these B-splines is completely contained in the parametric domain referred to as initial domain

$$\tilde{\mathcal{B}}^0 = \bigcup_{A \in \mathcal{I}^0} \operatorname{supp}(B^{0,A}) = \tilde{\mathcal{B}}. \tag{2.11}$$

As we have seen above, a single B-spline on level 0 can be replaced by its children, which correspond to B-splines defined on level 1

$$B_{\boldsymbol{p}}^{1,2\boldsymbol{i}-\boldsymbol{1}+\boldsymbol{j}}(\boldsymbol{\xi}) = \prod_{l=1}^{d} B_{p_l}^{0,i_l}(2\xi^l - j_l h_l^0 - \xi_{i_l}^{0,l}), \tag{2.12}$$

with bisected knot spans

$$\Xi_l^1 = \left[\xi_1^{1,l} = \xi_1^{0,l}, \xi_2^{1,l} = \frac{1}{2}(\xi_1^{0,l} + \xi_2^{0,l}), \xi_3^{1,l} = \xi_2^{0,l}, \ldots, \xi_{2(\mathrm{n}_l+p_l)+1}^{1,l} = \xi_{\mathrm{n}_l+p_l+1}^{0,l}\right]. \tag{2.13}$$

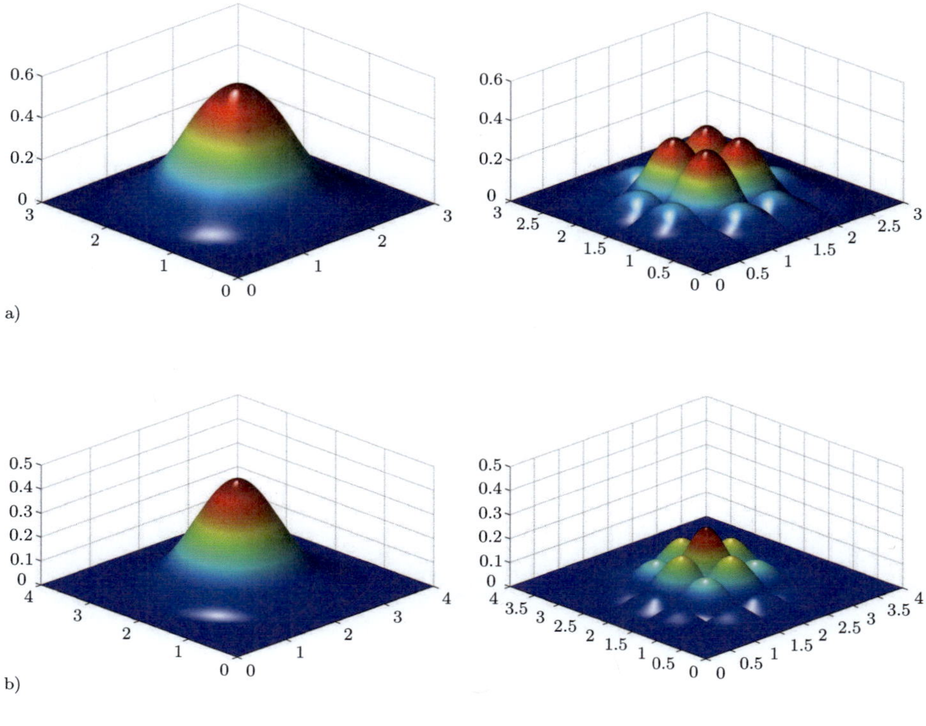

Figure 2.5: Subdivision of single B-spline basis functions in 2D. a) Quadratic B-spline with its scaled children. b) Cubic B-spline with its scaled children.

In practice, the recursive application of the refinement procedure leads to a multi level hierarchy such that we rewrite (2.10) and (2.12) as follows

$$B_p^{k,i}(\boldsymbol{\xi}) = \sum_{j=0}^{p+1} \bar{s}_{j+1} B_p^{k+1,2i-1+j}(\boldsymbol{\xi}), \tag{2.14}$$

where k is the level index and the subdivision vector $\bar{\boldsymbol{s}}$ includes the scaling information, cf. (2.10).

For an intuitive element wise construction of the refinement we select a set of elements on level k which constitutes the subdomain $\tilde{\mathcal{B}}^k_{\text{full}}$. Subsequently, we replace each B-spline which has support on $\tilde{\mathcal{B}}^k_{\text{full}}$ such that

$$\bigcup_{A \in \mathcal{I}^k \setminus \mathcal{J}^k} \text{supp}(B^{k,A}) \cap \tilde{\mathcal{B}}^k_{\text{full}} = \emptyset, \tag{2.15}$$

where $\mathcal{I}^k$ and $\mathcal{J}^k$ are the index sets of introduced and replaced level k B-splines, respectively. Now, the introduced B-splines on level $k+1$ are listed in the index set $\mathcal{I}^{k+1}$ and the union of their supports constitutes the domain

$$\tilde{\mathcal{B}}^{k+1} = \bigcup_{A \in \mathcal{I}^{k+1}} \text{supp}(B^{k+1,A}) = \bigcup_{A \in \mathcal{J}^k} \text{supp}(B^{k,A}). \tag{2.16}$$

Note that in general each B-spline contained in $\mathcal{I}^k$ can be replaced by B-splines of the next level. However, to maintain an intuitive construction and to avoid unnecessary complications we stipulate

$$\tilde{\mathcal{B}}^{k-1}_{\text{full}} \supseteq \tilde{\mathcal{B}}^k_{\text{full}}, \tag{2.17}$$

i.e. transition areas $\tilde{\mathcal{B}}^k \cap \tilde{\mathcal{B}}^{k-1}_{\text{full}}$ where the support is not purely provided by B-splines of level k are excluded from a full refinement to level $k+1$. In addition, this satisfies the non-intersection condition $\partial\tilde{\mathcal{B}}^k \cap \partial\tilde{\mathcal{B}}^{k+1} = \emptyset$ as required in Kraft [55] and leads to a nested sequence of domains

$$\tilde{\mathcal{B}}^0 \supsetneq \tilde{\mathcal{B}}^1 \supsetneq \ldots \tilde{\mathcal{B}}^k \supsetneq \tilde{\mathcal{B}}^{k+1} \supsetneq \ldots \supsetneq \tilde{\mathcal{B}}^{K_{\max}}, \tag{2.18}$$

where K_{max} denotes global maximum of refined levels. A further condition, which needs to be considered, is

$$\text{supp}^{\text{o}}(B^{k,A}) \not\subset \tilde{\mathcal{B}}^{k+1}_{\text{o}}, \tag{2.19}$$

for all $A \in \mathcal{I}^k \setminus \mathcal{J}^k$, i.e. the index set $\mathcal{J}^k$ has to be extended by level k B-splines whose open support is fully contained in $\tilde{\mathcal{B}}^{k+1}_{\text{o}}$ to avoid linear dependence of the hierarchical set of B-splines. Here $\tilde{\mathcal{B}}^{k+1}_{\text{o}}$ is defined in analogy to (2.16), except that we consider the open support instead of the support defined above. Such a situation may occur if $\tilde{\mathcal{B}}^k_{\text{full}}$ is not connected, whereas the transition areas consisting of level k and scaled level $k+1$ B-splines overlap as shown in Figure 2.6.

The subdivision approach given in (2.14) relates to the replacement of single B-splines. Concerning the replacement of an entire set of B-splines it is more convenient to utilize a vector-matrix notation

$$\boldsymbol{B}^k = \boldsymbol{S}^{k+1}\boldsymbol{B}^{k+1}, \tag{2.20}$$

where the vectors $\boldsymbol{B}^k$ and $\boldsymbol{B}^{k+1}$ contain removed B-splines on level k and introduced B-splines on level $k+1$, respectively, and $\boldsymbol{S}^{k+1}$ is the global subdivision matrix. Formally, we assume that (2.20) preserves the global numbering , i.e. $\boldsymbol{S}^{k+1}$ is a sparse matrix of size $((2^k(\mathfrak{n}_1+2)-2)\cdot\ldots\cdot(2^k(\mathfrak{n}_d+2)-2))\times((2^{k+1}(\mathfrak{n}_1+2)-2)\cdot\ldots\cdot(2^{k+1}(\mathfrak{n}_d+2)-2))$.

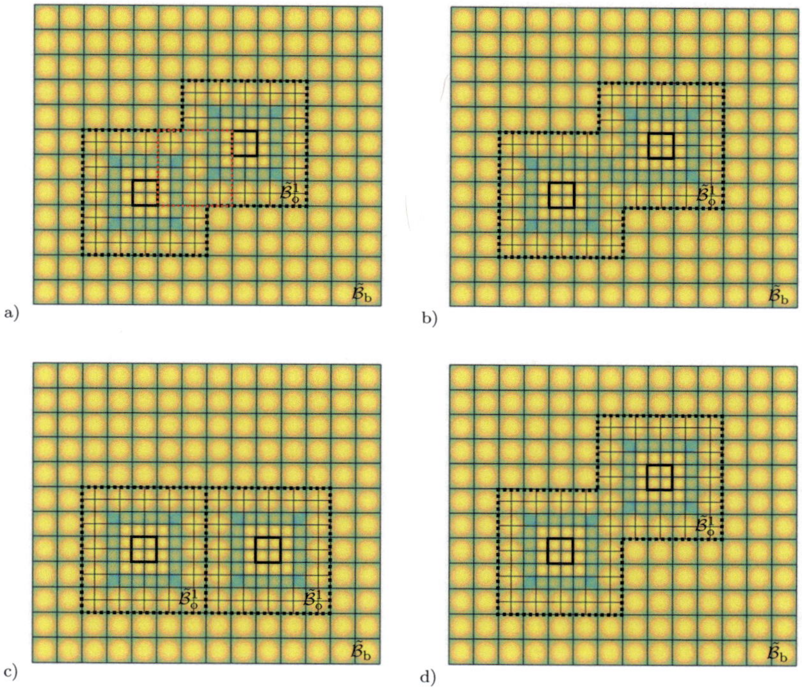

Figure 2.6: Local refinement of quadratic B-spline basis in 2D. The yellow point in the middle of each element represents the peak of a B-spline. In each of the four cases the subdomain $\tilde{\mathcal{B}}^0_{\text{full}}$ (solid box in black) is not connected and consists of two elements. a) The domain $\tilde{\mathcal{B}}^1_{\text{o}}$ is connected and the open support of a remaining level 0 B-spline (dashed box in red) is fully contained. b) Recovering of linear independence by replacing the level 0 B-spline. c) The domain $\tilde{\mathcal{B}}^1_{\text{o}}$ is not connected. d) The domain $\tilde{\mathcal{B}}^1_{\text{o}}$ is connected, but it exists no level 0 B-spline whose open support is fully contained.

Example

In order to clarify the proposed refinement procedure we discuss a one dimensional example in the following. For the set of quadratic B-splines shown in Figure 2.2 we select $\tilde{\mathcal{B}}^0_{\text{full}} = [4,6) \cup [11,12)$ as domain to be refined fully and thus $\mathcal{J}^0 = \{3,4,5,6,10,11,12\}$ as index set of B-splines to be replaced on level 0. As illustrated in Figure 2.7a and b, this satisfies the condition given by (2.15) and requires the introduction of level 1 B-splines contained in $\mathcal{I}^1 = \{5,6,7,8,9,10,11,12,13,14,19,20,21,22,23,24,25,26\}$. The associated subdivision matrix $\boldsymbol{S}^1$ is a rectangular matrix with $|\mathcal{J}^0| = 7$ non-zero rows and $|\mathcal{I}^1| = 18$ non-zero columns which are given by

$$\begin{bmatrix}
\frac{1}{4} & \frac{3}{4} & \frac{3}{4} & \frac{1}{4} & 0 & 0 & 0 & 0 & 0 & 0 & 0 & 0 & 0 & 0 & 0 & 0 & 0 & 0 \\
0 & 0 & \frac{1}{4} & \frac{3}{4} & \frac{3}{4} & \frac{1}{4} & 0 & 0 & 0 & 0 & 0 & 0 & 0 & 0 & 0 & 0 & 0 & 0 \\
0 & 0 & 0 & 0 & \frac{1}{4} & \frac{3}{4} & \frac{3}{4} & \frac{1}{4} & 0 & 0 & 0 & 0 & 0 & 0 & 0 & 0 & 0 & 0 \\
0 & 0 & 0 & 0 & 0 & 0 & \frac{1}{4} & \frac{3}{4} & \frac{3}{4} & \frac{1}{4} & 0 & 0 & 0 & 0 & 0 & 0 & 0 & 0 \\
0 & 0 & 0 & 0 & 0 & 0 & 0 & 0 & 0 & 0 & \frac{1}{4} & \frac{3}{4} & \frac{3}{4} & \frac{1}{4} & 0 & 0 & 0 & 0 \\
0 & 0 & 0 & 0 & 0 & 0 & 0 & 0 & 0 & 0 & 0 & 0 & \frac{1}{4} & \frac{3}{4} & \frac{3}{4} & \frac{1}{4} & 0 & 0 \\
0 & 0 & 0 & 0 & 0 & 0 & 0 & 0 & 0 & 0 & 0 & 0 & 0 & 0 & \frac{1}{4} & \frac{3}{4} & \frac{3}{4} & \frac{1}{4}
\end{bmatrix}. \tag{2.21}$$

Note that single level 1 B-splines are introduced multiply times within the hierarchical B-spline set as stipulated by (2.21). This leads to a linear dependence of the set of B-splines even if they are scaled differently, see Figure 2.7c. Hence, we do not use the subdivision matrix itself in practice, but rather we determine a scaling vector $s^1_j = \sum_i S^1_{i,j}$ which provides a unique summed scaling factor for each introduced B-spline. Here we obtain

$$\boldsymbol{s}^1 = \left[\ 0,0,0,0,\tfrac{1}{4},\tfrac{3}{4},1,1,1,1,1,1,\tfrac{3}{4},\tfrac{1}{4},0,\ldots,0,\tfrac{1}{4},\tfrac{3}{4},1,1,1,1,\tfrac{3}{4},\tfrac{1}{4},0,0,0,0\ \right]. \tag{2.22}$$

Now, the created hierarchical B-spline basis is linearly independent (see Kraft [55] for the proof on linear independence) and satisfies the partition of unity on $\tilde{\mathcal{B}}_b = [2,14)$ as indicated by the red line in Figure 2.7d. Moreover, we select $\tilde{\mathcal{B}}^1_{\text{full}} = [4,4.5) \cup [11.5,12)$ to continue the refinement. Thus, we obtain $\mathcal{J}^1 = \{7,8,9,22,23,24\}$ and $\mathcal{I}^2 = \{13,14,15,16,17,18,19,20,43,44,45,46,47,48,49,50\}$ as index set of B-splines to be replaced on level 1 and index set of B-splines to be introduced on level 2, respectively, see Figure 2.8a and b. Note that, due to (2.17), each B-spline which can be determined for $\mathcal{J}^1$ has a scaling factor equal to one. This greatly simplifies the construction of the scaling vector $\boldsymbol{s}^2$ and allows an efficient implementation. For the current example we obtain

$$\boldsymbol{s}^2 = \left[\ 0,\ldots,0,\tfrac{1}{4},\tfrac{3}{4},1,1,1,1,\tfrac{3}{4},\tfrac{1}{4},0,\ldots,0,\tfrac{1}{4},\tfrac{3}{4},1,1,1,1,\tfrac{3}{4},\tfrac{1}{4},0,\ldots,0\ \right], \tag{2.23}$$

whereas the hierarchical B-spline basis refined up to level 2 is depicted in Figure 2.8c. Finally, we apply an additional two level refinement to the left boundary of interval $\tilde{\mathcal{B}}_\text{b}$, i.e. we expand $\tilde{\mathcal{B}}^0_{\text{full}}$ by $[2,2.5)$ and $\tilde{\mathcal{B}}^1_{\text{full}}$ by $[2,2.25)$, respectively. The first refinement step necessitates the replacement of the level 0 B-splines $B^{0,1}$ and $B^{0,2}$. Note that $B^{0,3}$ was already being replaced within the former refinement, see Figure 2.7. Now, the refinement procedure demands the introduction of four not yet introduced level 1 B-splines as shown in Figure 2.9a. However, $B^{1,1}$ and $B^{1,2}$ have no support on $\tilde{\mathcal{B}}_b$ and the introduction of $B^{1,1}$ would violate the non-intersection condition. Hence, we do not add this B-splines to $\mathcal{I}^1$ and for the same reason we also omit the introduction of $B^{2,5}$ and $B^{2,6}$ within the second refinement step, see Figure 2.9b. Note that, due to (2.19), the B-spline $B^{1,6}$ needs also to be replaced since its open support is fully contained in $\tilde{\mathcal{B}}^2_\text{o}$, c.f. Figure 2.6. Eventually, the hierarchical B-spline basis with additional boundary refinement is given in Figure 2.9c.

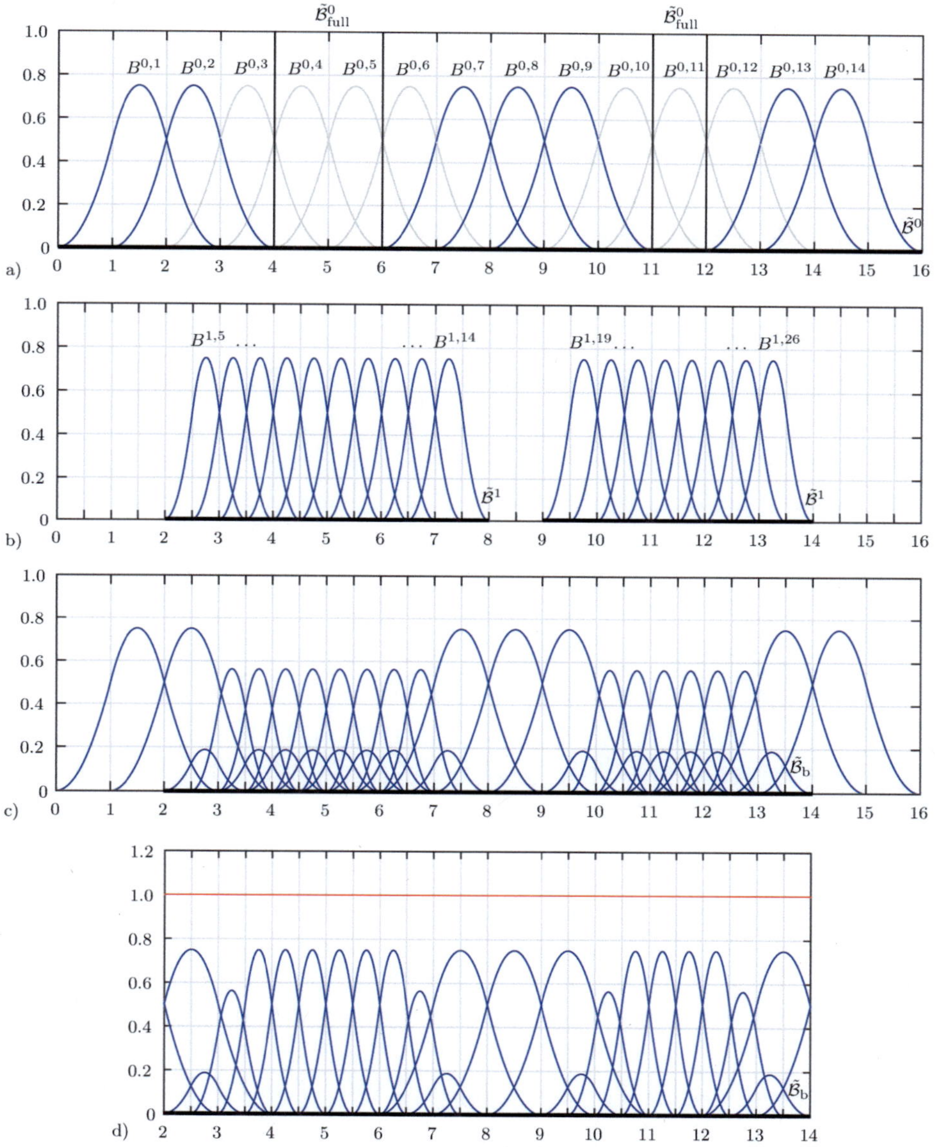

Figure 2.7: Local refinement of a quadratic B-spline basis in 1D. a) Removal of the seven level 0 B-splines in gray. b) Introduction of 18 B-splines on level 1. c) Hierarchical B-spline set. d) Hierarchical B-spline basis refined up to level 1 and partition of unity (red line).

Overall, we obtain $\mathcal{I}^1 = \{3,4,5,6,7,8,9,10,11,12,13,14,19,20,21,22,23,24,25,26\}$ and $\mathcal{J}^0 = \{1,2,3,4,5,6,10,11,12\}$ as index set of introduced and replaced B-splines, respectively, for the refinement up to level 1 along with the scaling vector

$$\boldsymbol{s}^1 = \left[\ 0,0,1,1,1,1,1,1,1,1,1,1,\tfrac{3}{4},\tfrac{1}{4},0,\ldots,0,\tfrac{1}{4},\tfrac{3}{4},1,1,1,1,\tfrac{3}{4},\tfrac{1}{4},0,0,0,0\ \right], \tag{2.24}$$

while $\mathcal{I}^2 = \{7,8,9,10,11,12,13,14,15,16,17,18,19,20,43,44,45,46,47,48,49,50\}$ and $\mathcal{J}^1 = \{3,4,5,6,7,8,9,22,23,24\}$ are the corresponding index sets for the refinement up to level 2 and

$$\boldsymbol{s}^2 = \left[\ 0,\ldots,0,1,1,1,1,1,1,1,1,1,1,1,1,\tfrac{3}{4},\tfrac{1}{4},0,\ldots,0,\tfrac{1}{4},\tfrac{3}{4},1,1,1,1,\tfrac{3}{4},\tfrac{1}{4},0,\ldots,0\ \right] \tag{2.25}$$

is the corresponding scaling vector.

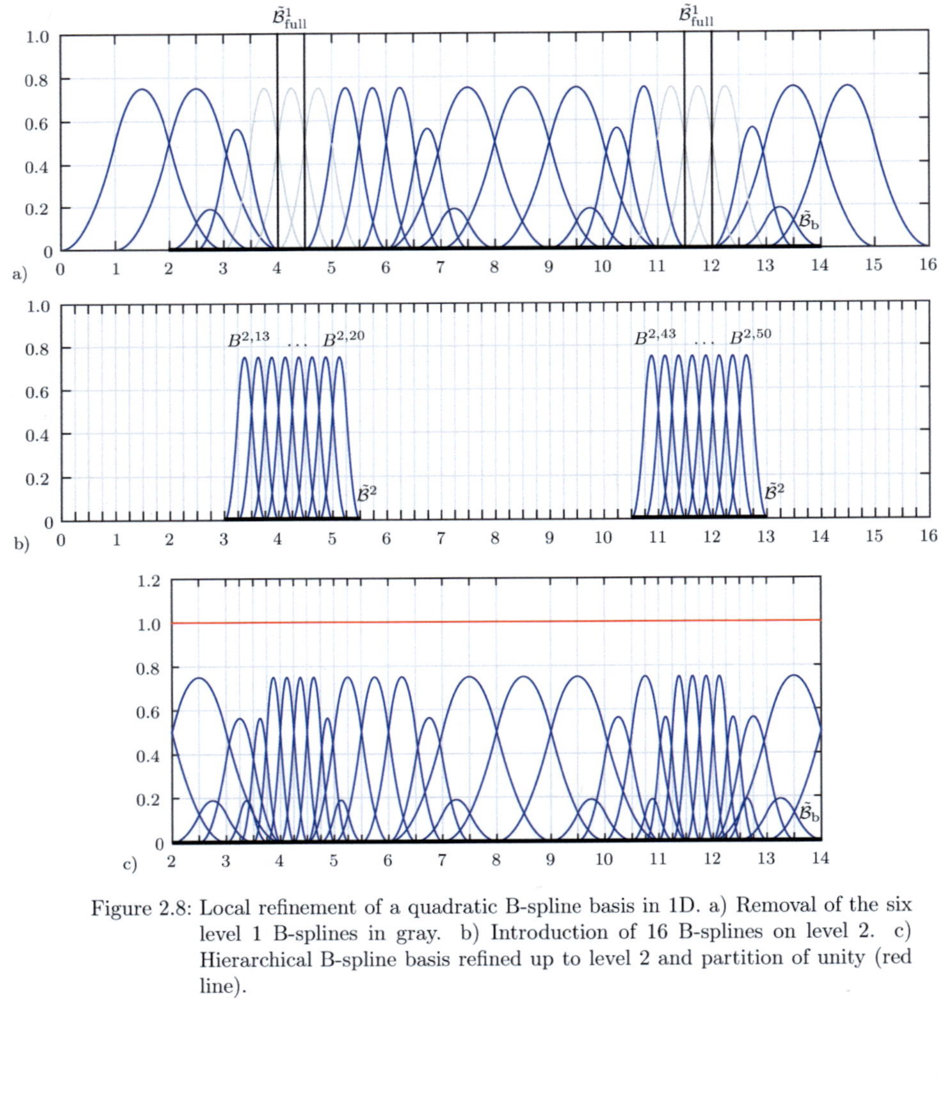

Figure 2.8: Local refinement of a quadratic B-spline basis in 1D. a) Removal of the six level 1 B-splines in gray. b) Introduction of 16 B-splines on level 2. c) Hierarchical B-spline basis refined up to level 2 and partition of unity (red line).

2.3 Geometric parametrization

Following the paradigm of the isogeometric concept the geometry is exactly represented and has to remain unchanged in case of refinements, i.e. the parametric domain $\tilde{\mathcal{B}}_{\mathrm{b}}$ introduced in section 2.1 correlates to the physical domain $\mathcal{B}$ independent of the existence of local refinements. The B-spline basis functions B^A are associated with a net of control points $\boldsymbol{X}_A \in \mathbb{R}^d$ such that a geometrical mapping $\mathfrak{F} : \tilde{\mathcal{B}}_{\mathrm{b}} \to \mathcal{B}$ can be defined to link the parameter and the physical space

$$\boldsymbol{X} = \mathfrak{F}(\boldsymbol{\xi}) = \sum_A B^A \boldsymbol{X}_A = \sum_{\boldsymbol{i}} B^{\boldsymbol{i}}_{\boldsymbol{p}}(\boldsymbol{\xi}) \boldsymbol{X}_{\boldsymbol{i}}, \tag{2.26}$$

cf. da Veiga et al. [22]. Moreover, we assume the existence of the inverse mapping $\mathfrak{F}^{-1} : \mathcal{B} \to \tilde{\mathcal{B}}_{\mathrm{b}}$ with same continuity as $\mathfrak{F}$.

Accounting for local refinements the control meshes as well as the associated primal variables (e.g. the actual position for solid mechanical problems, the temperature for thermal problems or the phase-field parameters for corresponding phase-field problems) need also to be created. For an arbitrary field $u(\boldsymbol{\xi})$ defined on level k follows

$$u(\boldsymbol{\xi}) = \boldsymbol{B}^k \cdot \boldsymbol{u}^k = (\boldsymbol{S}^{k+1} \boldsymbol{B}^{k+1}) \cdot \boldsymbol{u}^k = \boldsymbol{B}^{k+1} \cdot (\boldsymbol{S}^{k+1^T} \boldsymbol{u}^k), \tag{2.27}$$

where the vector $\boldsymbol{u}^k$ contains the control variables associated with the replaced B-splines in $\boldsymbol{B}^k$. From (2.27) we deduce that the new set of control variables is given via

$$\boldsymbol{u}^{k+1} = \boldsymbol{S}^{k+1^T} \boldsymbol{u}^k. \tag{2.28}$$

Obviously, the field $u(\boldsymbol{\xi})$ remains unchanged by using the level $k+1$ setting. However, by inserting (2.28) in (2.27) the subdivision matrix vanishes such that the B-spline basis loses its scaling informations. The approximation of the field $u(\boldsymbol{\xi})$ utilizing the hierarchical B-spline basis reads

$$u(\boldsymbol{\xi}) = \sum_{k=0}^{K_{\max}} \boldsymbol{B}^k_{\mathrm{b}} \cdot \boldsymbol{u}^k_{\mathrm{b}} = \sum_{k=0}^{K_{\max}} \sum_{A \in \mathcal{I}^k \setminus \mathcal{J}^k} s^k_A B^{k,A} \frac{u^k_A}{s^k_A}, \tag{2.29}$$

where $\boldsymbol{B}^k_{\mathrm{b}}$ and $\boldsymbol{u}^k_{\mathrm{b}}$ are the vectors of scaled B-splines and adapted control variables contained in the index set $\mathcal{I}^k \setminus \mathcal{J}^k$, respectively. Note that the B-splines defined on level 0 remain unscaled, i.e. $s^0_A = 1 \, \forall A \in \mathcal{I}^0 \setminus \mathcal{J}^0$, and note also that the solution field itself is not constrained by (2.28) such that the B-spline basis is able to represent a field with finer details.

An adaptive hierarchical refinement approach is beyond the scope of this work, but can be found in Jiang & Dolbow [52]. A least square fitting based on the rectangular subdivision matrix can be applied for coarsening.

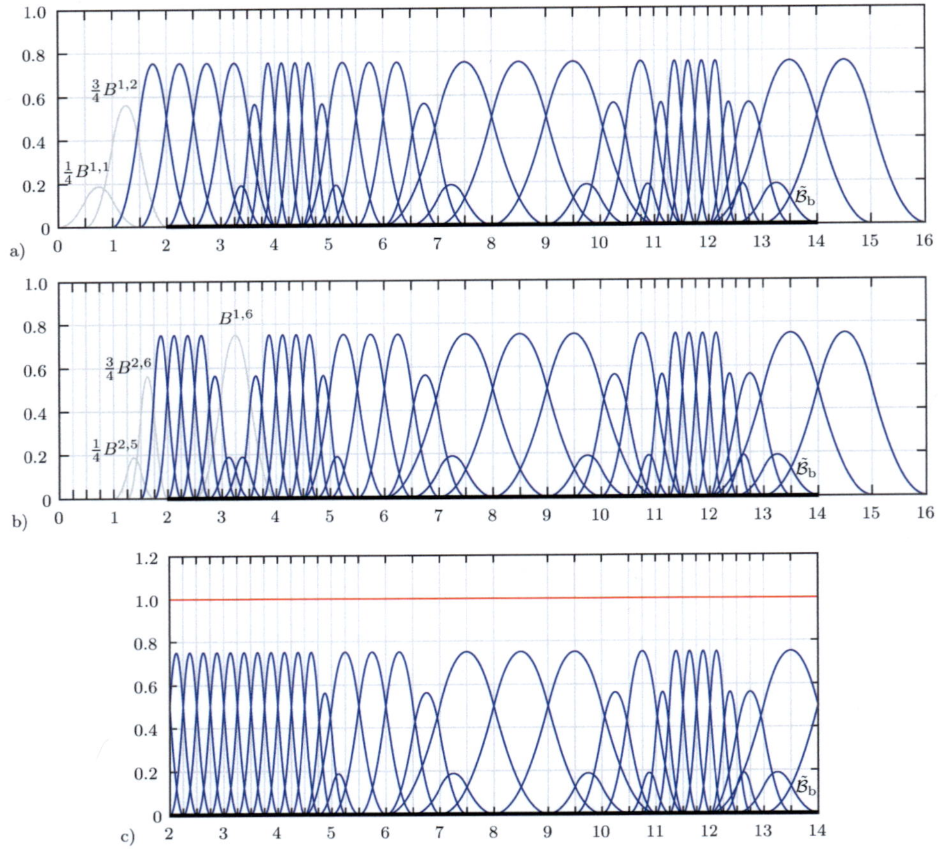

Figure 2.9: Local refinement of a quadratic B-spline basis in 1D. a) Replacement of two level 0 B-splines through two level 1 B-splines at the left boundary. B-splines in gray will not be introduced. b) Replacement of four level 1 B-splines through six level 2 B-splines at the left boundary. B-splines in gray will not be introduced and will be replaced, respectively. c) Hierarchical B-spline basis with additional boundary refinement up to level 2 and partition of unity (red line). The mesh is indicated by the vertical lines in gray.

The mesh of parametric domain $\tilde{\mathcal{B}}_\mathrm{b}$ is constituted by a set of elements from multiple levels, see, e.g. Figure 2.9. For the construction of the mesh we introduce a piecewise constant scalar function

$$k_{\max}(\boldsymbol{\xi}) = \max_k\{0 \leq k \leq K_{\max} | \boldsymbol{\xi} \in \operatorname{supp}(B^{k,A}),\ A \in \mathcal{I}^k \setminus \mathcal{J}^k\} \tag{2.30}$$

which specifies the local maximum refinement level at $\boldsymbol{\xi}$, cf. Bornemann and Cirak [13]. Using this function, we define the index set of elements

$$\tilde{\mathcal{T}}^k = \{\boldsymbol{j} \in \mathbb{N}^d | \tilde{\mathcal{B}}_{k,\boldsymbol{j}} \subseteq \tilde{\mathcal{B}}_\mathrm{b} \quad \text{and} \quad k_{\max}(\tilde{\mathcal{B}}_{k,\boldsymbol{j}}) = k\} \tag{2.31}$$

with which we create a non-overlapping mesh such that[5]

$$\tilde{\mathcal{B}}_{\mathrm{b}} = \bigcup_{\substack{j \in \tilde{\mathcal{T}}^k \\ k \geq 0}} \tilde{\mathcal{B}}_{k,j} = \bigcup_{e \in \mathbb{E}} \tilde{\mathcal{B}}_e. \tag{2.32}$$

Evaluation

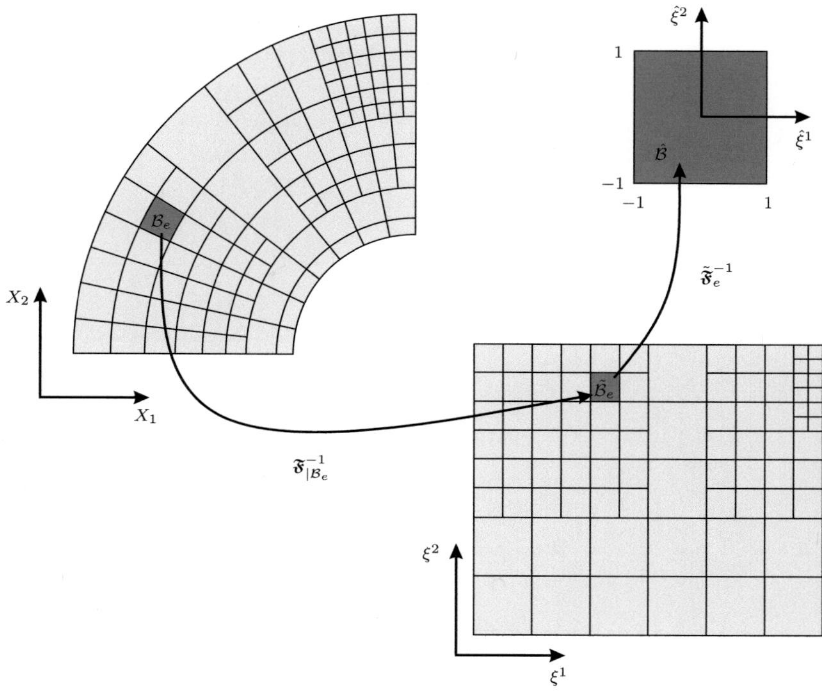

Figure 2.10: Pull back of a physical element onto the element in the parametric domain via the inverse geometrical mapping and subsequently onto the parent element via an inverse affine mapping.

The numerical evaluation of an integral over $\mathcal{B}_e$ resulting from a specific weak formulation is performed by using Gaussian quadrature. For this purpose, we pull back the element contribution onto the element in the parametric domain via the inverse mapping $\mathfrak{F}^{-1}_{|\mathcal{B}_e} : \mathcal{B}_e \rightarrow \tilde{\mathcal{B}}_e$ and subsequently onto the bi unit parent element through a further inverse mapping $\tilde{\mathfrak{F}}_e^{-1} : \tilde{\mathcal{B}}_e \rightarrow \hat{\mathcal{B}}$. As shown in Figure 2.10, the coordinates in the physical space and in the parameter space are given by $\boldsymbol{X} = [X_1, \ldots, X_d]$ and $\boldsymbol{\xi} = [\xi^1, \ldots, \xi^d]$, respectively, whereas we introduce $\hat{\boldsymbol{\xi}} = [\hat{\xi}^1, \ldots, \hat{\xi}^d]$ as the coordinates in the parent element.

[5] The index e relates to the global numbering of elements of the mesh in the parameter space as well as in the physical space.

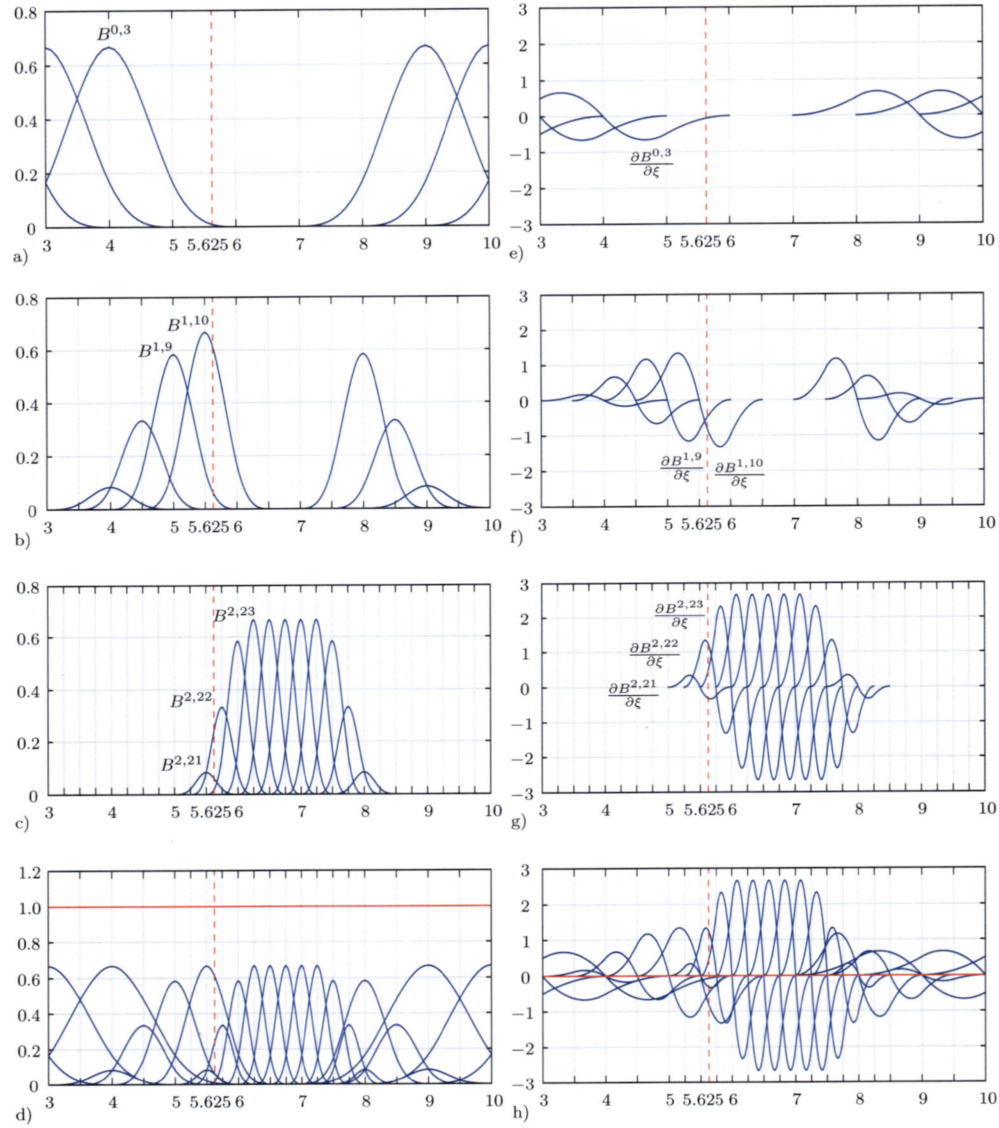

Figure 2.11: Evaluation of a univariate hierarchical B-spline basis of order $p = 3$ at $\xi = 5.625$ (dotted line in red). a) One B-spline basis function evaluated on level 0. b) Two B-spline basis functions evaluated on level 1. c) Three B-spline basis functions evaluated on level 2 d) In total six B-spline basis functions have support at $\xi = 5.625$. e) to g) Evaluation of corresponding derivatives with respect to the parametric coordinate ξ. h) First derivative of the hierarchical B-spline basis with respect to ξ and sum of the derivatives (red line).

The mapping $\tilde{\mathfrak{F}}_e$ is an affine transformation and reads

$$\boldsymbol{\xi} = \tilde{\mathfrak{F}}_e(\hat{\boldsymbol{\xi}}) = \frac{1}{2}\begin{bmatrix} (\tilde{\xi}^{k_e,1}_{j_1+1} - \tilde{\xi}^{k_e,1}_{j_1})\hat{\xi}^1 + (\tilde{\xi}^{k_e,1}_{j_1+1} + \tilde{\xi}^{k_e,1}_{j_1}) \\ \vdots \\ (\tilde{\xi}^{k_e,d}_{j_d+1} - \tilde{\xi}^{k_e,d}_{j_d})\hat{\xi}^d + (\tilde{\xi}^{k_e,d}_{j_d+1} + \tilde{\xi}^{k_e,d}_{j_d}) \end{bmatrix}, \tag{2.33}$$

where $k_e = k_{\max}(\tilde{\mathcal{B}}_e)$. Now, we evaluate the basis functions which have support in the eth element as well as their derivatives with respect to the parametric coordinates at each quadrature point. Note that on elements lying in transition areas, B-spline basis functions originally defined on lower levels than k_e have support as well. Nevertheless, we always apply (2.33) to calculate the evaluation point in the parametric domain, see Figure 2.11. Another point to note is that the number of basis functions and thus the number of degrees of freedom per element are not constant such that the size of element matrices and vectors varies. This has to be handled flexibly within the FEA framework. To obtain the gradient of a basis function $\nabla B^{k,A}$ we apply the chain rule as follows

$$\frac{\partial B^{k,A}}{\partial X_l} = \frac{\partial B^{k,A}}{\partial \xi^i}\frac{\partial \xi^i}{\partial X_l}, \tag{2.34}$$

where $\partial\boldsymbol{\xi}/\partial\boldsymbol{X}$ is the inverse gradient of the mapping $\mathfrak{F}_{|\mathcal{B}_e}$. Moreover, the Laplacian of a basis function $\Delta B^{k,A}$ is given in straightforward manner by using the second derivative

$$\frac{\partial^2 B^{k,A}}{\partial X_l \partial X_l} = \frac{\partial^2 B^{k,A}}{\partial \xi^i \partial \xi^j}\frac{\partial \xi^i}{\partial X_l}\frac{\partial \xi^j}{\partial X_l} + \frac{\partial B^{k,A}}{\partial \xi^i}\frac{\partial^2 \xi^i}{\partial X_l \partial X_l}. \tag{2.35}$$

The Laplacian $\Delta B^{k,A}$ is required to FEA treatment of fourth order partial differential equations (e.g. the Cahn-Hilliard equation or the Kuramoto-Sivashinsky equation). Eventually, to perform the numerical integration in the parent element we need also to calculate the Jacobian determinant of the composition of both mappings $\mathfrak{F}_{|\tilde{\mathcal{B}}_e} \circ \tilde{\mathfrak{F}}_e$ at each quadrature point. The Jacobian determinant reads

$$\hat{J} = \det\left(\frac{\partial X_l}{\partial \xi^i}\frac{\partial \xi^i}{\partial \hat{\xi}^j}\right). \tag{2.36}$$

2.4 Hierarchical refinement of NURBS

The subdivision approach given by (2.10) is only valid for uniform B-spline bases and has to be modified for non-uniform B-spline bases. To be specific, the entries of the subdivision vector $\bar{\boldsymbol{s}}$ are no longer given as usual binomial coefficients and thus they need to be determined for each replaced B-spline, see, e.g. Figure 2.12. For the subdivision of a univariate B-spline on level k we can obtain the subdivision vector by solving a system of linear equations

$$\begin{bmatrix} B_p^{k+1,2i-1}(\bar{\xi}_1) & & & \\ B_p^{k+1,2i-1}(\bar{\xi}_2) & B_p^{k+1,2i}(\bar{\xi}_2) & & 0 \\ \vdots & \vdots & \ddots & \\ B_p^{k+1,2i-1}(\bar{\xi}_{p+2}) & B_p^{k+1,2i}(\bar{\xi}_{p+2}) & \dots & B_p^{k+1,2i+p+1}(\bar{\xi}_{p+2}) \end{bmatrix} \begin{bmatrix} \bar{s}_1 \\ \bar{s}_2 \\ \vdots \\ \bar{s}_{p+2} \end{bmatrix} = \begin{bmatrix} B_p^{k,i}(\bar{\xi}_1) \\ B_p^{k,i}(\bar{\xi}_2) \\ \vdots \\ B_p^{k,i}(\bar{\xi}_{p+2}) \end{bmatrix}, \tag{2.37}$$

where $\bar{\xi}_j = \frac{1}{2}(\tilde{\xi}^{k+1}_{2i-2+j} + \tilde{\xi}^{k+1}_{2i-1+j})$ with $j \in \{1, 2, \dots, p+2\}$. Note that other choices for $\bar{\xi}_j$ are possible, however, by using the proposed setting we ensure that the created system of equations has a unique solution regardless of the polynomial order of the B-spline. This is evident in (2.37), where the matrix is a lower triangular matrix with non-zero entries on the main diagonal.

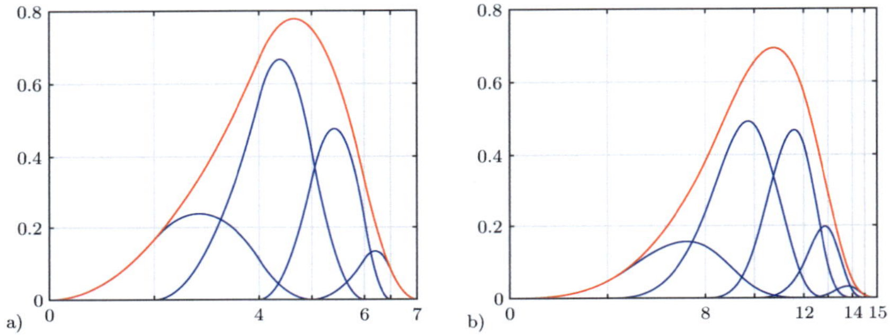

Figure 2.12: Subdivision of single B-splines defined by non-uniform knot sequences in 1D. a) $\bar{\boldsymbol{s}} = [1/3, 5/6, 2/3, 1/6]$ is applied to subdivide a quadratic B-spline defined by $\Xi = [0, 4, 6, 7]$ and b) $\bar{\boldsymbol{s}} = [5/21, 5/7, 5/7, 2/7, 1/21]$ is applied to subdivide a cubic B-spline defined by $\Xi = [0, 8, 12, 14, 15]$.

The situation renders more complex in the present of repeated knots as often used for NURBS. In addition to the construction of the scaling vector we have to note that also the construction of the new knot vector given by (2.13) is invalid. However, in analogy to (2.13), the modified knot vector of level $k+1$ can be obtained via

$$\tilde{\xi}^{k+1}_j = \begin{cases} \tilde{\xi}^k_i & \text{if } j = 2i-1 \\ \frac{1}{2}(\tilde{\xi}^k_i + \tilde{\xi}^k_{i+1}) & \text{if } j = 2i \end{cases}, \tag{2.38}$$

whereas we create the vector of knot multiplicities via

$$m^{k+1}_j = \begin{cases} m^k_i & \text{if } j = 2i-1 \\ 1 & \text{if } j = 2i \end{cases}. \tag{2.39}$$

The knot vector of level $k+1$ follows then in a straightforward manner by using the modified knot vector along with the vector of knot multiplicities such that we bisect only knot spans whose width is different from zero. Note that the relation $\xi^k_i = \xi^{k+1}_{2i-1}$ and thus the specification of B-splines to be introduced in (2.14) is not valid anymore. Moreover, due to the repeated knots it is insufficient to consider only the beginning position of a B-spline for its identification. In particular, we have to identify a B-spline to be introduced through its entire support. Formally, for the replacement of a single B-spline $B^{k,A}$ the index set of required level $k+1$ B-splines is given as follows

$$\mathcal{I}_A^{k+1} = \{B \in \mathbb{N} | \text{supp}(B^{k+1,B}) \subseteq \text{supp}(B^{k,A})\}. \tag{2.40}$$

Then, the entries of the subdivision vector $\bar{\boldsymbol{s}}$ can be achieved by solving a system of equations similar to that in 2.37. Furthermore, in Hesch et al. [44] we provide a construction rule for modified subdivision matrices to deal with repeated knots at domain boundaries. Figure 2.13 illustrates the subdivision of different univariate B-splines defined by knot sequences including repeated knots. The extension to the multivariate case follows in a straightforward manner by taking the tensor product structure of B-splines into account and the generalization of (2.14) is formulated as

$$B^{k,A} = \sum_{B \in \mathcal{I}_A^{k+1}} s_B B^{k+1,B}. \tag{2.41}$$

Following the extension outlined in (2.7) the replacement of a single NURBS basis function defined on level k reads

$$R^{k,A} = \frac{\sum\limits_{B \in \mathcal{I}_A^{k+1}} s_B B^{k+1,B} w_B^{k+1}}{W}, \tag{2.42}$$

where W is the weight function. From geometrical point of view, a NURBS geometry in d dimensional space is obtained by the projection of a $d+1$ dimensional B-spline geometry, see Cottrell et al. [18]. Accounting for refinements we project back a given NURBS geometry defined on level k into the B-spline space where

$$\bar{\boldsymbol{X}}_A^k = [X_A^{k,1} w_A^k, \ldots, X_A^{k,d} w_A^k, w_A^k] \tag{2.43}$$

is the associated control net. As mentioned above, the geometry has to be preserved as the mesh is refined, i.e. we calculate the control net of level $k+1$ via

$$\bar{\boldsymbol{X}}^{k+1} = \boldsymbol{S}^{k+1^T} \bar{\boldsymbol{X}}^k \tag{2.44}$$

such that

$$\boldsymbol{B}^k \bar{\boldsymbol{X}}^k = \boldsymbol{B}^{k+1} \bar{\boldsymbol{X}}^{k+1}, \tag{2.45}$$

cf. (2.27) and (2.28), respectively. This implies that also the weight function remains unchanged. Afterwards, we project the refined control net to the d dimensional NURBS space. Formally, this is analogous to standard knot insertion technique for NURBS. Note that, in practice, we make use of the hierarchical B-spline basis to calculate the weight function

$$W(\boldsymbol{\xi}) = \sum_{k=0}^{K_{\max}} \boldsymbol{B}_{\mathrm{b}}^k \boldsymbol{w}_{\mathrm{b}}^k, \tag{2.46}$$

cf. (2.29).

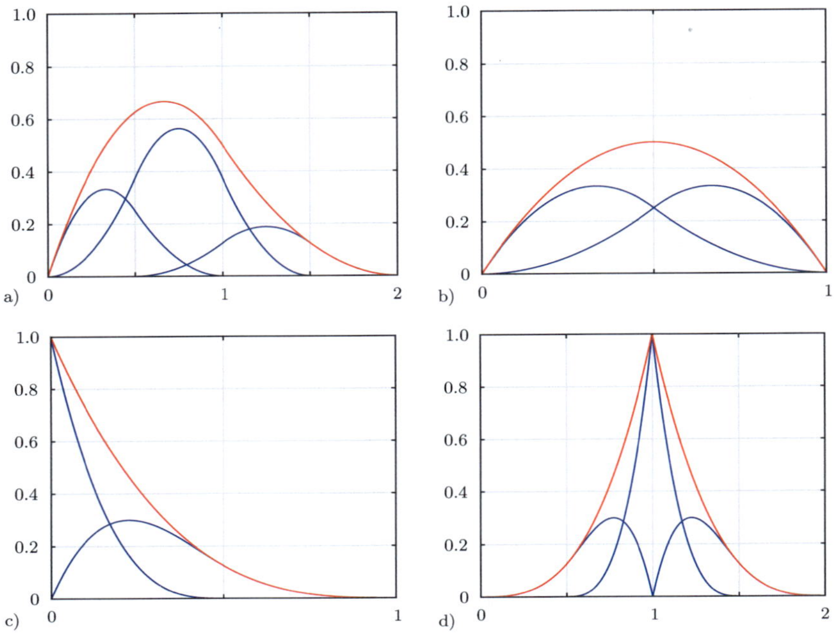

Figure 2.13: Subdivision of single B-splines defined by knot sequences with repeated knots in 1D. a) $\bar{\boldsymbol{s}} = [1/2, 3/4, 1/4]$ is applied to subdivide a quadratic B-spline defined by $\Xi = [0, 0, 1, 2]$, b) $\bar{\boldsymbol{s}} = [1/2, 1/2]$ is applied to subdivide a quadratic B-spline defined by $\Xi = [0, 0, 1, 1]$, c) $\bar{\boldsymbol{s}} = [1, 1/2]$ is applied to subdivide a cubic B-spline defined by $\Xi = [0, 0, 0, 0, 1]$ and d) $\bar{\boldsymbol{s}} = [1/2, 1, 1/2]$ is applied to subdivide a cubic B-spline defined by $\Xi = [0, 1, 1, 1, 2]$.

2.5 Numerical verification

In this section we demonstrate the investigation of a physical problem which requires for approximation at least C^1-continuity, cf. Hesch et al. [44]. Starting with a one dimensional modification of the Kuramoto-Sivashinsky model, we derive an analytical solution (see Xu & Shu [100]) and compare this reference solution with simulation results obtained for different local refinements. Subsequently, the two dimensional case will demonstrate the consistency of the chosen approach with local refinements.

Kuramoto-Sivashinsky problem

The Kuramoto-Sivashinsky model has been proposed for the simulation of different thermodynamical systems far away from equilibrium state, e.g. fluctuations in fluid films and instabilities in laminar flame fronts, see, Paniconi & Elder [73], among many others. This transient fourth-order evolution equation accounts for the physics of arising chaotic instabilities in the spatial as well as the temporal domain.

The numerical resolution of the fourth order operator is traditionally realized using finite difference schemes. In the context of FEA for irreducible systems, global C^1-continuity is required, see Gomez & Paris [33] for an application of NURBS basis functions.

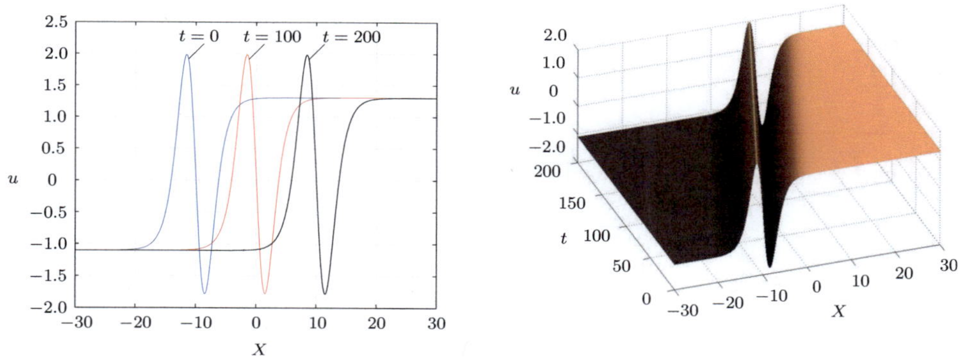

Figure 2.14: Analytical solution $u(X,t)$ of the one dimensional Kuramoto-Sivashinsky equation over position X and time t.

Assume a Lipschitz bounded domain $\mathcal{B} \subseteq \mathbb{R}^d$ in the time interval $\mathcal{T} = [0, T]$. Then the Kuramoto-Sivashinsky problem is stated as

$$\dot{u} + \nu\Delta^2(u) + \Delta(u) + \alpha u - \|\nabla(u)\|^2 = 0, \tag{2.47}$$

with boundary and initial conditions, given as

$$\nabla u \cdot \boldsymbol{n} = 0, \quad \nabla(u + \Delta u) \cdot \boldsymbol{n} = 0, \tag{2.48}$$

and

$$u(\boldsymbol{X}, 0) = u_0(\boldsymbol{X}), \tag{2.49}$$

respectively. Therein $\dot{u}$ is the derivative with respect to time, whereas the constitutive parameters ν and α are used to model diffusion induced chaos and dissipative structures in reaction diffusion systems. Note, that (2.48) is often replaced by periodic boundary conditions, see, e.g. Gomez & Paris [33].

Next we introduce the functional space of admissible test functions

$$\mathcal{V}^u = \{\delta u \in \mathcal{H}^2(\mathcal{B}) |\ \nabla\delta u = 0 \text{ on } \partial\mathcal{B} \times \mathcal{T}\}. \tag{2.50}$$

The weak form of the problem reads

$$\int\limits_{\mathcal{B}} \delta u \dot{u} \ \mathrm{d}V + \int\limits_{\mathcal{B}} \nu\Delta(\delta u)\Delta(u) \ \mathrm{d}V - \int\limits_{\mathcal{B}} \nabla(\delta u) \cdot \nabla(u) \ \mathrm{d}V + \int\limits_{\mathcal{B}} \delta u \left(\alpha u - \|\nabla(u)\|^2\right) \mathrm{d}V = 0, \tag{2.51}$$

for all $\delta u \in \mathcal{V}^u$. For the temporal discretization, we subdivide the time interval $\mathcal{T}$ into a sequence of times $t_0, \dots, t_n, t_{n+1}, \dots, T$ and assume that the state at t_n, denoted by $u_{B,n}$, is known. Then we approximate the state at t_{n+1} with time step size $\Delta t = t_{n+1} - t_n$ using a second order accurate mid-point evaluation. The corresponding fully discrete system takes the form[6]

$$\delta u_A \Bigg[\int_{\mathcal{B}} R^A R^B \, \mathrm{d}V \; \frac{u_{B,n+1} - u_{B,n}}{\Delta t} + \int_{\mathcal{B}} \nu \Delta R^A \Delta R^B \, \mathrm{d}V \; u_{B,n+\frac{1}{2}} - \int_{\mathcal{B}} \nabla R^A \cdot \nabla R^B \, \mathrm{d}V \; u_{B,n+\frac{1}{2}} + \int_{\mathcal{B}} R^A (\alpha R^B u_{B,n+\frac{1}{2}} - \|\nabla R^B u_{B,n+\frac{1}{2}}\|^2) \, \mathrm{d}V \Bigg] = 0. \tag{2.52}$$

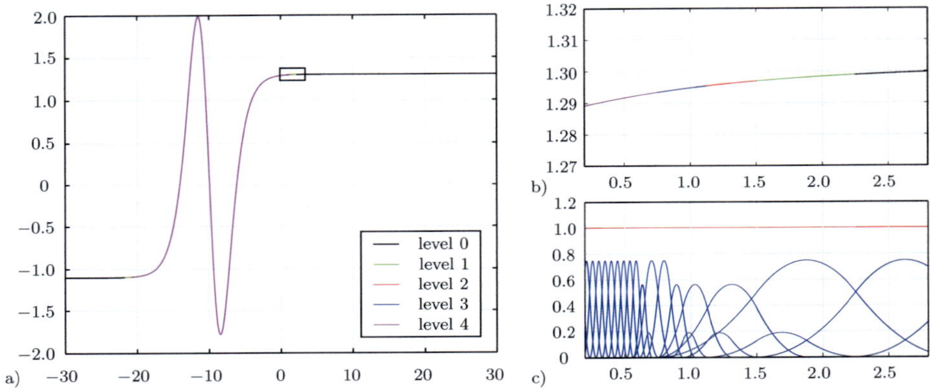

Figure 2.15: One dimensional Kuramoto-Sivashinsky problem. a) Local refinement up to level 4 plotted along with the initial configuration. b) Transition area transition from level zero to level four. c) Shape functions as well as their sum in the transition area.

In a first step, a slightly simplified one dimensional model presented in detail in Anders et al. [3] is analyzed. Let the domain $\mathcal{B} = [\mathcal{L}_0, \mathcal{L}] \subseteq \mathbb{R}$ be defined in the time interval $\mathcal{T}$. The corresponding weak form reads

$$\int_{\mathcal{B}} \delta u \dot{u} \, \mathrm{d}V + \int_{\mathcal{B}} \frac{\partial^2 \delta u}{\partial X^2} \frac{\partial^2 u}{\partial X^2} \, \mathrm{d}V - \int_{\mathcal{B}} \frac{\partial \delta u}{\partial X} \frac{\partial u}{\partial X} \, \mathrm{d}V - \frac{1}{2} \int_{\mathcal{B}} \frac{\partial \delta u}{\partial X} u^2 \, \mathrm{d}V = 0, \tag{2.53}$$

for all $\delta u \in \mathcal{V}^u$. The analytical solution for $\mathcal{B} = [-30, 30]$ is given by

$$u(X,t) = c + \frac{15}{19}\sqrt{\frac{11}{19}}(-9\tanh(k(X - ct - X_0)) + 11\tanh^3(k(X - ct - X_0))), \tag{2.54}$$

[6] Here and in the following we neglect the summation over the level index k for readability and use R for NURBS basis functions as well as for B-spline basis functions.

with parameters $c = 0.1$, $k = \frac{1}{2}\sqrt{\frac{11}{19}}$ and $X_0 = -10$, see Figure 2.14. The boundary conditions read $u(-30,t) = g_1(t)$ and $u(30,t) = g_2(t)$, whereas the initial condition is given by $u(X,0) = u_0(X)$. The corresponding Parameters g_1, g_2 and u_0 can be calculated from (2.53) and (2.54).

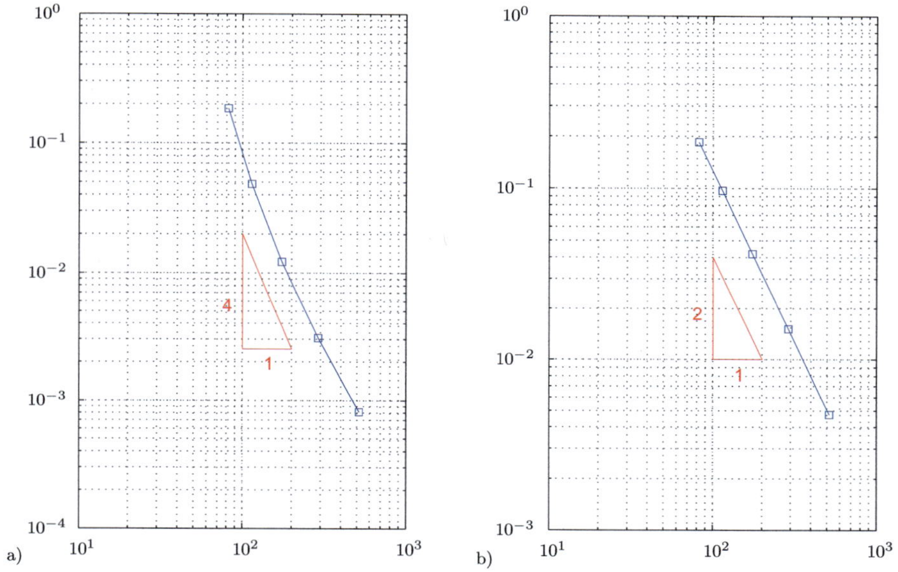

Figure 2.16: Convergence plot, L^2-error over the number of degrees of freedom. a) Convergence for local hierarchical refinement. b) Convergence for uniform h-refinement.

In order to illustrate the capabilities of the hierarchical refinement concept we evaluate the deviation of the numerical solution related to the reference solution. To be specific, numerical results using a constant time step size $\Delta t = 0.005$ are compared with analytical results at time $t = 2$. To ensure consistency of the higher order approach, we apply first a uniform global h-refinement scheme. Then, we refine the region passed by the shock wave within the time interval $\mathcal{T} = [0,2]$ level-by-level up to the fourth level, see Figure 2.15 for details. The obtained results are plotted in Figure 2.16, the displayed error is measured in the L^2-norm $\|u - u^h\|_2$. For the global h-refinement the expected second order convergence is obtained, see Figure 2.16b. For the hierarchical refinement strategy we achieved higher rates of convergence, see Figure 2.16a. The corresponding condition numbers of the stiffness matrix for both refinements are presented in Table 2.1.

Next, we present the numerical results of the two dimensional version of the Kuramoto-Sivashinsky problem (2.47) on $\mathcal{B} = [0,100]^2$. Following the investigations in Gomez & Paris [33] we apply periodic boundary conditions and a uniform random perturbation between $[-0.05, 0.05]$ of the homogeneous field $u = 0$ as initial condition, shown in Figure

2.17a. Furthermore, we set the first constitutive parameter $\nu = 1$, whereas the settings of the second parameter are $\alpha = 0.225$, $\alpha = 0.210$ and $\alpha = 0.195$, respectively, to obtain the three basic states found by Paniconi & Elder [73]. Here the hexagonal state ($\alpha = 0.225$) yields a spatially ordered stationary solution, the breathing hexagonal state ($\alpha = 0.210$) is characterized by a periodic oscillation of hexagonal cells and for the disordered state ($\alpha = 0.195$) the solution shows a chaotic behavior. The original mesh consists of 128×128 elements. A refined area of 20×20 elements for level 1 is located at the center of the domain to demonstrate the validity of the refinement technique for multivariate problems of higher order, see Figure 2.17b. Finally, we show the numerical solution for the three basic states of the Kuramoto-Sivashinsky problem in Figure 2.18 for each case to four discrete points in time.

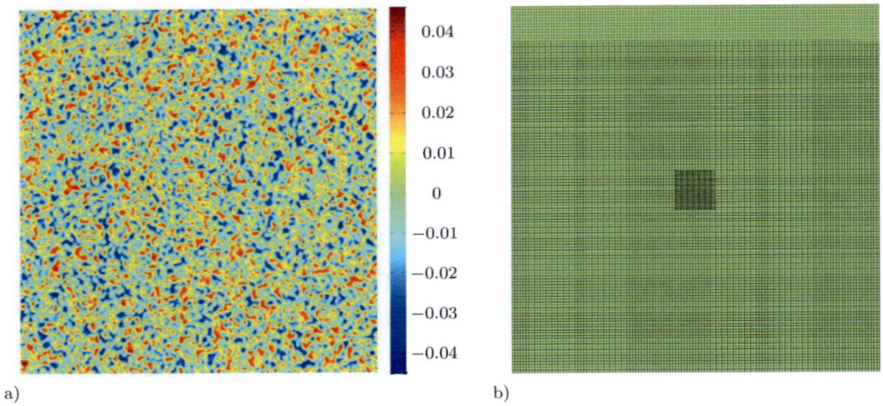

Figure 2.17: Two dimensional Kuramoto-Sivashinsky problem. a) Initial configuration of the Kuramoto-Sivashinsky problem. b) Computational mesh with local refinement in the middle.

Remark 1: We observe for $\alpha = 0.225$ that the time until the system achieved a perfect stationary solution is in general larger compared to the usage of a uniform 128×128 elements mesh. In a series of numerical test we observed that using a finer background mesh yields for both, the refined as well as the unrefined case, nearly identical results.

Table 2.1: Condition number of stiffness matrix for uniform h-refinement and hierarchical refinement.

Number of degrees of freedom / (hierarchical levels)	Uniform h-refinement	Hierarchical refinement
82 / (0)	57.3	57.3
114 / (1)	43.9	200.2
174 / (2)	20.3	213.8
290 / (3)	18.6	1388.7
516 / (4)	89.5	11106.0

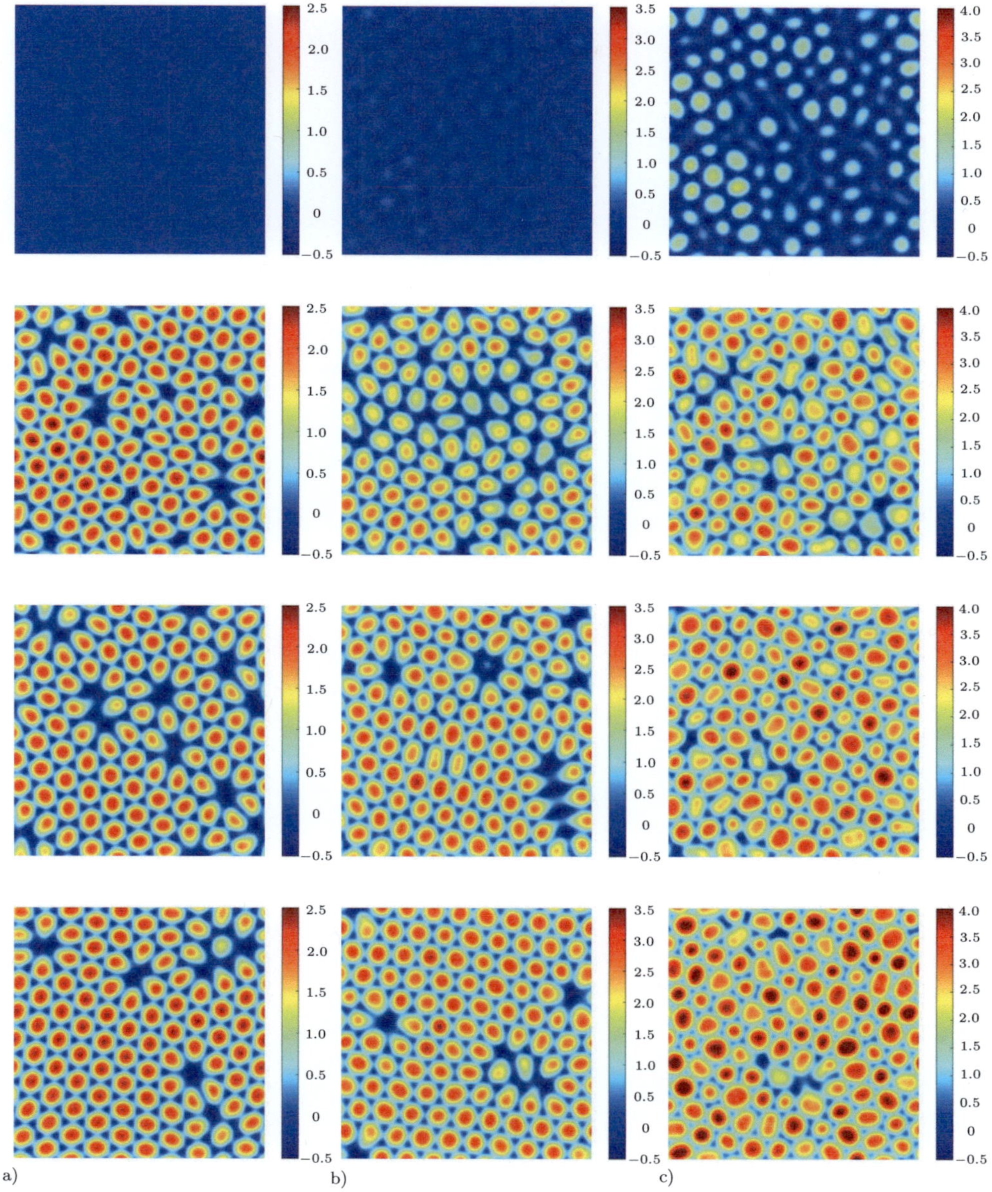

Figure 2.18: Numerical solution of the two dimensional Kuramoto-Sivashinsky problem at times $t = [100, 500, 2500, 12500]$ (from top to bottom). a) Hexagonal state with $\alpha = 0.225$. b) Breathing hexagonal state with $\alpha = 0.210$. c) Disordered state with $\alpha = 0.195$.

3 Large strain thermoelasticity

In this chapter we present a novel computational framework for large strain thermoelasticity. The ideas of a new formulation for polyconvex large strain elasticity originally introduced by Schröder et al. [82] and recently extended by Bonet et al. [10] are employed to non-linear coupled thermoelasticity, see textbooks like Holzapfel [46] and Malvern [60] for a comprehensive survey of this topic. In particular, we introduce the deformation gradient (line map), its co-factor (area map) and its determinant (volume map) along with the absolute temperature as independent variables to formulate a free Helmholtz energy density function. Moreover, we introduce work conjugate stresses to the extended kinematic set to define a complementary energy principle of the Hellinger–Reissner type, where the new conjugate stresses along with the deformed geometry and the absolute temperature are primal variables. Classical stress tensors, e.g. the first Piola-Kirchhoff stress tensor, relate in elegant manner to the newly introduced conjugate stresses by using the cross product between second order tensors. The finite element discretization as proposed in Bonet et al. [10] relies on a quadratic approximation of the geometry and a discontinuous linear and constant interpolation of the conjugate stresses using tetrahedral elements. This discretization scheme is adapted to the isogeometric concept as introduced in the previous chapter and a standard condensation procedure is employed to reduce the global system related to the stress unknowns. Eventually, quasi-static as well as transient numerical examples are investigated to demonstrate the capability of the proposed framework.

3.1 Configuration and kinematics

We consider a three dimensional thermoelastic body, i.e. $d = 3$, in its reference configuration occupying a domain $\mathcal{B}_0$ with boundary $\partial\mathcal{B}_0$, see Figure 3.1. Assuming that $\mathcal{B}$ denotes the current configuration we introduce a smooth non-linear deformation mapping

$$\boldsymbol{\varphi}(\boldsymbol{X}, t) : \mathcal{B}_0 \times \mathcal{T} \rightarrow \mathbb{R}^d, \tag{3.1}$$

to map a material point $\boldsymbol{X}$ in its reference configuration to its current position $\boldsymbol{x} = \boldsymbol{\varphi}(\boldsymbol{X}, t)$ at time $t \in \mathcal{T} = [0, T]$, where $T \in \mathbb{R}^+$. Furthermore, we introduce the absolute temperature

$$\theta(\boldsymbol{X}, t) : \mathcal{B}_0 \times \mathcal{T} \rightarrow \mathbb{R}_{\geq 0}, \tag{3.2}$$

which is assumed to be a sufficiently smooth function. The unknowns $\{\boldsymbol{\varphi}, \theta\}$ from a configuration space in $\mathbb{R}^{d+1}$ represent the primal degrees of freedom to be found for all times of interest.

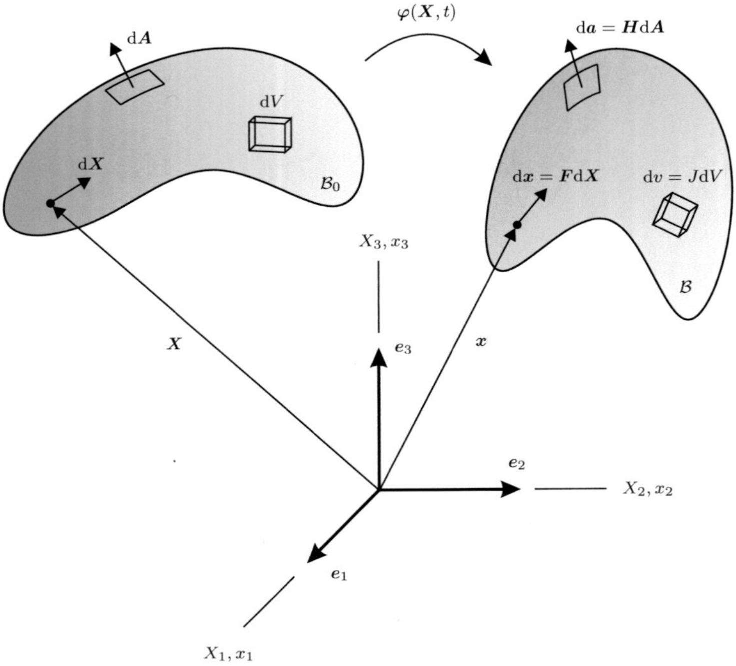

Figure 3.1: Deformation mapping of a continuum body from a reference configuration $\mathcal{B}_0$ into a current configuration $\mathcal{B}$ and associated strain measures $\boldsymbol{F}, \boldsymbol{H}$ and J.

The deformation gradient tensor $\boldsymbol{F} : \mathcal{B}_0 \times \mathcal{T} \to \mathbb{R}^{d\times d}$ is commonly defined as material gradient of the current configuration

$$\boldsymbol{F} = \nabla(\varphi) = \frac{\partial \varphi}{\partial \boldsymbol{X}}. \tag{3.3}$$

Moreover, we introduce the determinant $J : \mathcal{B}_0 \times \mathcal{T} \to \mathbb{R}$ as well as the co-factor $\boldsymbol{H} : \mathcal{B}_0 \times \mathcal{T} \to \mathbb{R}^{d\times d}$ of the deformation gradient, usually defined by $J = \det(\boldsymbol{F})$ and $\boldsymbol{H} = J\boldsymbol{F}^{-1}$, respectively. As shown in Figure 3.1, the three strain measures $\boldsymbol{F}, \boldsymbol{H}$ and J map differential line, area and volume elements between the reference and the current configuration, i.e. $\mathrm{d}\boldsymbol{x} = \boldsymbol{F}\mathrm{d}\boldsymbol{X}$, $\mathrm{d}\boldsymbol{a} = \boldsymbol{H}\mathrm{d}\boldsymbol{A}$ and $\mathrm{d}v = J\mathrm{d}V$. Regarding the latter both mappings, we provide an alternative representation by using the cross product between second order tensors[1]. Thus, the co-factor or area map tensor is given as

$$\boldsymbol{H} = \operatorname{cof}(\boldsymbol{F}) = \frac{1}{2}(\boldsymbol{F} \times \boldsymbol{F}) \tag{3.4}$$

and the Jacobian determinant or volume map reads

$$J = \det(\boldsymbol{F}) = \frac{1}{6}(\boldsymbol{F} \times \boldsymbol{F}) : \boldsymbol{F}. \tag{3.5}$$

[1] The definition of the tensor cross product operation is given in Appendix A.

Note that the tensor cross product operation was originally introduced by de Boer [23] and for the first time applied in the context of solid mechanics in Bonet et al. [10]. The usage of the tensor cross product operation simplifies tremendously the first and second directional derivatives of (3.4) and (3.5) with respect to virtual and incremental variations $\delta\boldsymbol{\varphi}$ and $\Delta\boldsymbol{\varphi}$, since differentiation of the inverse of the deformation gradient is not required. For the co-factor follows immediately

$$\mathrm{D}\boldsymbol{H}[\delta\boldsymbol{\varphi}] = \boldsymbol{F} \times \nabla(\delta\boldsymbol{\varphi}), \quad \mathrm{D}^2\boldsymbol{H}[\delta\boldsymbol{\varphi}, \Delta\boldsymbol{\varphi}] = \nabla(\delta\boldsymbol{\varphi}) \times \nabla(\Delta\boldsymbol{\varphi}), \tag{3.6}$$

whereas the derivatives of the determinant of the deformation gradient read

$$\mathrm{D}J[\delta\boldsymbol{\varphi}] = \boldsymbol{H} : \nabla(\delta\boldsymbol{\varphi}), \quad \mathrm{D}^2 J[\delta\boldsymbol{\varphi}, \Delta\boldsymbol{\varphi}] = \boldsymbol{F} : (\nabla(\delta\boldsymbol{\varphi}) \times \nabla(\Delta\boldsymbol{\varphi})). \tag{3.7}$$

3.2 First and second law of thermodynamics

The first law of thermodynamics states that the energy of a system remains preserved during an entire thermodynamic process. Although, the energy can be transformed from one form of energy into another, it can neither be created nor destroyed. This means that the temporal change of the total energy, which consists of the kinetic energy T and the internal energy E, has to be equal to the sum of the external mechanical power P^{ext} and the heat supply Q, i.e.

$$\dot{T}(t) + \dot{E}(t) = P^{\mathrm{ext}}(t) + Q(t). \tag{3.8}$$

In material description the global energy balance law (3.8) can be written as

$$\frac{\mathrm{d}}{\mathrm{d}t} \int_{\mathcal{B}_0} \left(\frac{1}{2} \rho_0 \dot{\boldsymbol{\varphi}} \cdot \dot{\boldsymbol{\varphi}} + e \right) \mathrm{d}V = \int_{\partial\mathcal{B}_0} (\boldsymbol{T} \cdot \dot{\boldsymbol{\varphi}} + Q_{\mathrm{N}}) \mathrm{d}A + \int_{\mathcal{B}_0} (\boldsymbol{B} \cdot \dot{\boldsymbol{\varphi}} + R) \mathrm{d}V, \tag{3.9}$$

where ρ_0 and e are the mass density and the internal energy density defined with respect to the reference volume. In addition, $\boldsymbol{T}$ denotes the material description of applied surface tractions and $\boldsymbol{B}$ the material description of prescribed body forces, whereas Q_{N} is the normal heat flux and R is the heat source, measured per unit reference surface area and reference volume, respectively. Applying Cauchy's stress theorem $\boldsymbol{T} = \boldsymbol{P}\boldsymbol{N}$ and Stokes heat flux theorem $Q_{\mathrm{N}} = -\boldsymbol{Q} \cdot \boldsymbol{N}$ along with some standard calculations yields

$$\int_{\mathcal{B}_0} \dot{e} \mathrm{d}V = \int_{\mathcal{B}_0} (\boldsymbol{P} : \dot{\boldsymbol{F}} + \underline{(\mathrm{Div}(\boldsymbol{P}) + \boldsymbol{B} - \rho_0 \ddot{\boldsymbol{\varphi}})} \cdot \dot{\boldsymbol{\varphi}} - \mathrm{Div}(\boldsymbol{Q}) + R) \mathrm{d}V, \tag{3.10}$$

where $\boldsymbol{P}$ and $\boldsymbol{Q}$ are the first Piola-Kirchhoff stress tensor and the Piola-Kirchhoff heat flux vector, respectively, whereas $\boldsymbol{N}$ denotes the outward unit normal vector to $\partial\mathcal{B}_0$. Note that the underlined vector-valued expression in (3.10) corresponds to the local form of the balance of linear momentum and is required to be equal to zero, such that we can express (3.9) in the more explicit form

$$\int_{\mathcal{B}_0} \dot{e} \mathrm{d}V = \int_{\mathcal{B}_0} (\boldsymbol{P} : \dot{\boldsymbol{F}} - \mathrm{Div}(\boldsymbol{Q}) + R) \mathrm{d}V. \tag{3.11}$$

Moreover, satisfaction of the balance of angular momentum leads to the symmetry condition $\boldsymbol{P}\boldsymbol{F}^{\mathrm{T}} = \boldsymbol{F}\boldsymbol{P}^{\mathrm{T}}$. Assuming that the volume of the reference configuration is arbitrary, we can deduce the local form of the balance of energy in material description as follows

$$\dot{e} = \boldsymbol{P} : \dot{\boldsymbol{F}} - \mathrm{Div}(\boldsymbol{Q}) + R. \tag{3.12}$$

As mentioned above, the first law of thermodynamics governs the transformation from one type of energy into another, maintaining the energy balance, but it does not make restrictions on the direction of the process. In the sense of the Clausius statement [17], heat never flows spontaneously from a colder to a warmer region without external work being performed on the system. To this end, we postulate the second law of thermodynamics which specifies limits on the direction of thermodynamic processes. To be specific, it states that the total entropy production Γ, determined as difference between the temporal change of entropy $\dot{S}$ and rate of entropy input $\tilde{Q}$, has to be greater than or equal to zero for all admissible thermodynamic processes

$$\Gamma(t) = \dot{S}(t) - \tilde{Q}(t) \geq 0. \tag{3.13}$$

In material description this inequality reads

$$\frac{\mathrm{d}}{\mathrm{d}t}\int\limits_{\mathcal{B}_0} \eta \mathrm{d}V + \int\limits_{\partial\mathcal{B}_0} \tilde{\boldsymbol{Q}} \cdot \boldsymbol{N} \mathrm{d}A - \int\limits_{\mathcal{B}_0} \tilde{R} \mathrm{d}V \geq 0, \tag{3.14}$$

where η is the entropy density defined with respect to the reference volume, which can be considered as a quantitative measure of microscopic randomness and disorder. In addition, $\tilde{\boldsymbol{Q}}$ denotes the Piola-Kirchhoff entropy flux vector and $\tilde{R}$ is the entropy source, measured per unit reference surface area and reference volume, respectively. Assuming that the entropy flux and the entropy source are related to the heat flux and the heat source by proportional factor $1/\theta$, i.e. $\tilde{\boldsymbol{Q}} = \boldsymbol{Q}/\theta$ and $\tilde{R} = R/\theta$, respectively, we obtain after some standard calculations

$$\int\limits_{\mathcal{B}_0} \left(\dot{\eta} + \frac{1}{\theta} \mathrm{Div}(\boldsymbol{Q}) - \frac{1}{\theta^2} \boldsymbol{Q} \cdot \nabla(\theta) - \frac{1}{\theta} R \right) \mathrm{d}V \geq 0, \tag{3.15}$$

cf. Hutter [50]. Analogous to (3.12), the local form of the inequality, known as Clausius-Duhem inequality, in the material configuration is given via

$$\theta\dot{\eta} + \mathrm{Div}(\boldsymbol{Q}) - \frac{1}{\theta}\boldsymbol{Q} \cdot \nabla(\theta) - R \geq 0. \tag{3.16}$$

The third term in (3.16) represents the dissipation by heat conduction

$$D^{\mathrm{con}} = -\frac{1}{\theta}\boldsymbol{Q} \cdot \nabla(\theta) \geq 0, \tag{3.17}$$

where the identity holds if the temperature gradient vanishes. (3.17) shows that heat must always flow against a temperature gradient. A further contribution to the total entropy production arises due to internal dissipation $D^{\text{int}} \geq 0$, see Simo [84]. In the case of perfect thermoelasticity, the internal dissipation is omitted and the total dissipation is given via the dissipation by heat conduction, i.e. $D = D^{\text{con}}$, such that

$$\theta \dot{\eta} = -\operatorname{Div}(\boldsymbol{Q}) + R, \tag{3.18}$$

which represents the local form of the balance of energy in entropy form for reversible processes ($D^{\text{int}} = 0$). Insertion of (3.18) into (3.12) leads to an alternative formulation, namely

$$\dot{e} = \boldsymbol{P} : \dot{\boldsymbol{F}} + \theta \dot{\eta}. \tag{3.19}$$

Furthermore, we introduce the free Helmholtz energy density via the Legendre transformation $\Psi = e - \theta\eta$ and rewrite (3.19) as follows

$$\dot{\Psi} = \boldsymbol{P} : \dot{\boldsymbol{F}} - \eta \dot{\theta}. \tag{3.20}$$

3.3 Constitutive theory

For the closure of the thermodynamical problem outlined in the previous section a constitutive model must be introduced. In particular, we define the free Helmholtz energy density as an at least twice continuously differentiable function of the deformation gradient and the absolute temperature

$$\Psi = \Psi(\boldsymbol{F}, \theta), \tag{3.21}$$

where we restrict ourselves to homogeneous material behavior. Knowing the constitutive relation (3.21), we may deduce directly thermodynamically consistent expressions for the first Piola-Kirchhoff stress tensor and the entropy from (3.20). By means of the Gibbs relation, which is obtained by determining the total time derivative of the free Helmholtz energy density function, (3.20) reads

$$0 = \left(\frac{\partial \Psi(\boldsymbol{F}, \theta)}{\partial \boldsymbol{F}} - \boldsymbol{P} \right) : \dot{\boldsymbol{F}} + \left(\frac{\partial \Psi(\boldsymbol{F}, \theta)}{\partial \theta} + \eta \right) \dot{\theta} \tag{3.22}$$

and for arbitrary choices of $\dot{\boldsymbol{F}}$ and $\dot{\theta}$ we obtain

$$\boldsymbol{P} = \boldsymbol{P}(\boldsymbol{F}, \theta) = \frac{\partial \Psi(\boldsymbol{F}, \theta)}{\partial \boldsymbol{F}} \quad \text{and} \quad \eta = \eta(\boldsymbol{F}, \theta) = -\frac{\partial \Psi(\boldsymbol{F}, \theta)}{\partial \theta}. \tag{3.23}$$

In order to complete the description of thermoelastic material behavior, (3.21) and (3.23) must be supplemented by a thermodynamically consistent constitutive relation for the Piola-Kirchhoff heat flux vector $\boldsymbol{Q}$. We define the heat flux vector as a function of the deformation gradient, the temperature and its gradient, i.e.

$$\boldsymbol{Q} = \boldsymbol{Q}(\boldsymbol{F}, \theta, \nabla(\theta)), \tag{3.24}$$

satisfying the inequality of the dissipation by heat conduction given in (3.17).

Note that the compatibility with the balance laws as well as the consistency with the second law represent an important constitutive principle which has to be satisfied to avoid the derivation of non-physical models, for a survey of constitutive principles see Altenbach & Altenbach [2].

3.3.1 Free Helmholtz energy

In order to specify an admissible free Helmholtz energy density function we consider a further constitutive principle. The principle of material frame indifference implies that Ψ has to be independent of the rotational components of $\boldsymbol{F}$. This is usually achieved by formulating Ψ as a function depending on $\boldsymbol{F}$ via the symmetric tensor $\boldsymbol{C} = \boldsymbol{F}^{\mathrm{T}}\boldsymbol{F}$ denoted as right Cauchy-Green tensor. Concerning isotropic material behavior, the dependency can be further simplified by utilizing the invariants of $\boldsymbol{C}$ defined as

$$I_C = \boldsymbol{F} : \boldsymbol{F}, \qquad II_C = \boldsymbol{H} : \boldsymbol{H} \qquad \text{and} \qquad III_C = J^2. \tag{3.25}$$

Thus, the free Helmholtz energy density function can be expressed through the three strain measures introduced in section 3.1 and temperature

$$\Psi(\boldsymbol{F}, \theta) = \tilde{\Psi}(\boldsymbol{F}, \boldsymbol{H}, J, \theta). \tag{3.26}$$

Moreover, it is often convenient to decouple the deviatoric and volumetric response of the material. Hence, we decompose the purely elastic part of the free Helmholtz energy density function into isochoric and volumetric components, Ψ^{iso} and Ψ^{vol}, respectively, as follows

$$\begin{aligned} \Psi(\boldsymbol{F}, \theta) &= \bar{\Psi}(\bar{\boldsymbol{F}}, \bar{\boldsymbol{H}}, J, \theta) \\ &= \bar{\Psi}^{\mathrm{iso}}(\bar{\boldsymbol{F}}, \bar{\boldsymbol{H}}) + \bar{\Psi}^{\mathrm{vol}}(J) + \bar{\Psi}^{\mathrm{the}}(\theta) + \bar{\Psi}^{\mathrm{cpl}}(J, \theta), \end{aligned} \tag{3.27}$$

where $\bar{\boldsymbol{F}}(\boldsymbol{F}, J) = J^{-1/3}\boldsymbol{F}$ and $\bar{\boldsymbol{H}}(\boldsymbol{H}, J) = J^{-2/3}\boldsymbol{H}$ denote the isochoric parts of the deformation gradient and its co-factor, respectively. In addition, the function $\bar{\Psi}^{\mathrm{the}}$ represents the potential for the purely thermal entropy, whereas the function $\bar{\Psi}^{\mathrm{cpl}}$ is the potential for the associated elastic structural entropy. Note that the latter one describes the thermomechanical coupling due to thermal expansion which can be modeled in the sense of the linear theory as

$$\bar{\Psi}^{\mathrm{cpl}}(J, \theta) = -d\gamma(\theta - \theta_{\mathrm{ref}}) \frac{\partial \bar{\Psi}^{\mathrm{vol}}(J)}{\partial J}, \tag{3.28}$$

where γ is the linear thermal expansion coefficient and θ_{ref} is the reference temperature, see Simo & Miehe [85]. Thus, a typical thermoelastic compressible Mooney-Rivlin material can be described by the free Helmholtz energy density function

$$\begin{aligned} \bar{\Psi}(\bar{\boldsymbol{F}}, \bar{\boldsymbol{H}}, J, \theta) = &\underbrace{\alpha(\bar{\boldsymbol{F}} : \bar{\boldsymbol{F}} - d) + \beta((\bar{\boldsymbol{H}} : \bar{\boldsymbol{H}})^{3/2} - d^{3/2})}_{\bar{\Psi}^{\mathrm{iso}}} + \underbrace{\frac{\kappa}{2}(J-1)^2}_{\bar{\Psi}^{\mathrm{vol}}} \\ &+ \underbrace{c\left(\theta - \theta_{\mathrm{ref}} - \theta \ln\left(\frac{\theta}{\theta_{\mathrm{ref}}}\right)\right)}_{\bar{\Psi}^{\mathrm{the}}} + \underbrace{(-d\gamma(\theta - \theta_{\mathrm{ref}})\kappa(J-1))}_{\bar{\Psi}^{\mathrm{cpl}}}, \end{aligned} \tag{3.29}$$

where α and β are positive material parameters related to the shear modulus and κ is the positive bulk modulus. Furthermore, the thermal material parameter $c \geq 0$ denotes the specific heat capacity which is assumed to be constant. The free Helmholtz energy density function given by (3.29) is polyconvex with respect to $\boldsymbol{F}$, $\boldsymbol{H}$ and J. Convexity is additive, a proof on polyconvexity of $\bar{\Psi}^{\text{iso}}$ and $\bar{\Psi}^{\text{vol}}$ can be found in Schröder & Neff [81] and the reference therein, whereas $\bar{\Psi}^{\text{cpl}}$ is linear in J. Note that for strain energy functions describing purely elastic material behavior in the large strain regime, polyconvexity is a fundamental mathematical requirement.

Remark 2: Regarding the Legendre transformation in section 3.3.4 the free Helmholtz energy density function given by (3.29) is constructed as simple as possible, however, the function do not satisfies the growth condition for $J \to 0_+$. In case of strong compression, we replace $\bar{\Psi}^{\text{vol}}$ in (3.29) via $\frac{\kappa}{4}((J-1)^2 + (\ln(J))^2)$, see e.g. Doll & Schweizerhof [28] for the development of volumetric strain energy functions.

3.3.2 Conjugate stresses, entropy and Hessian operator

The extended kinematic set along with the temperature $\{\boldsymbol{F}, \boldsymbol{H}, J, \theta\}$ enables us to introduce an associated set of work conjugate variables $\{\boldsymbol{\Sigma}_F, \boldsymbol{\Sigma}_H, \Sigma_J, \eta\}$ defined as

$$
\begin{aligned}
\boldsymbol{\Sigma}_F &= \frac{\partial \bar{\Psi}^{\text{iso}}}{\partial \bar{\boldsymbol{F}}} \frac{\partial \bar{\boldsymbol{F}}}{\partial \boldsymbol{F}} = J^{-1/3} \bar{\boldsymbol{\Sigma}}_F, \\
\boldsymbol{\Sigma}_H &= \frac{\partial \bar{\Psi}^{\text{iso}}}{\partial \bar{\boldsymbol{H}}} \frac{\partial \bar{\boldsymbol{H}}}{\partial \boldsymbol{H}} = J^{-2/3} \bar{\boldsymbol{\Sigma}}_H, \\
\Sigma_J &= \frac{\partial \bar{\Psi}^{\text{iso}}}{\partial \bar{\boldsymbol{F}}} : \frac{\partial \bar{\boldsymbol{F}}}{\partial J} + \frac{\partial \bar{\Psi}^{\text{iso}}}{\partial \bar{\boldsymbol{H}}} : \frac{\partial \bar{\boldsymbol{H}}}{\partial J} + \frac{\partial (\bar{\Psi}^{\text{vol}} + \bar{\Psi}^{\text{cpl}})}{\partial J} \\
&= -\frac{1}{3} J^{-4/3} \bar{\boldsymbol{\Sigma}}_F : \boldsymbol{F} - \frac{2}{3} J^{-5/3} \bar{\boldsymbol{\Sigma}}_H : \boldsymbol{H} + p, \\
\eta &= -\frac{\partial (\bar{\Psi}^{\text{the}} + \bar{\Psi}^{\text{cpl}})}{\partial \theta}.
\end{aligned}
\tag{3.30}
$$

Note that $\{\bar{\boldsymbol{\Sigma}}_F, \bar{\boldsymbol{\Sigma}}_H\}$ are work conjugate stresses to the isochoric strain components $\{\bar{\boldsymbol{F}}, \bar{\boldsymbol{H}}\}$. Accordingly, p denotes the work conjugate hydrostatic pressure to J which is positive in tension and negative in compression.

Next, we derive the relationship between the first Piola-Kirchhoff stress tensor $\boldsymbol{P}$ and the introduced sets of stresses $\{\boldsymbol{\Sigma}_F, \boldsymbol{\Sigma}_H, \Sigma_J\}$ and $\{\bar{\boldsymbol{\Sigma}}_F, \bar{\boldsymbol{\Sigma}}_H, p\}$, respectively. The variation of both energy functions in $(3.27)_1$ with respect to the primal variables formally reads

$$
\mathrm{D}\Psi[\,\mathrm{D}\boldsymbol{F}[\delta\boldsymbol{\varphi}], \delta\theta] = \mathrm{D}\bar{\Psi}[\,\mathrm{D}\boldsymbol{F}[\delta\boldsymbol{\varphi}], \mathrm{D}\boldsymbol{H}[\delta\boldsymbol{\varphi}], \mathrm{D}J[\delta\boldsymbol{\varphi}], \delta\theta]. \tag{3.31}
$$

Recalling (3.6) and (3.7) yields

$$
\begin{aligned}
\boldsymbol{P} : \nabla(\delta\boldsymbol{\varphi}) - \eta\delta\theta &= \boldsymbol{\Sigma}_{\boldsymbol{F}} : \mathrm{D}\boldsymbol{F}[\delta\boldsymbol{\varphi}] + \boldsymbol{\Sigma}_{\boldsymbol{H}} : \mathrm{D}\boldsymbol{H}[\delta\boldsymbol{\varphi}] + \Sigma_J\, \mathrm{D}J[\delta\boldsymbol{\varphi}] - \eta\delta\theta \\
&= (\boldsymbol{\Sigma}_{\boldsymbol{F}} + \boldsymbol{\Sigma}_{\boldsymbol{H}} \times \boldsymbol{F} + \Sigma_J \boldsymbol{H}) : \nabla(\delta\boldsymbol{\varphi}) - \eta\delta\theta.
\end{aligned}
\tag{3.32}
$$

Thus, we find that

$$\begin{aligned}\boldsymbol{P} &= \boldsymbol{\Sigma}_F + \boldsymbol{\Sigma}_H \times \boldsymbol{F} + \Sigma_J \boldsymbol{H} \\ &= J^{-1/3}\bar{\boldsymbol{\Sigma}}_F + J^{-2/3}\bar{\boldsymbol{\Sigma}}_H \times \boldsymbol{F} + \left(p - \frac{1}{3}J^{-4/3}\bar{\boldsymbol{\Sigma}}_F : \boldsymbol{F} - \frac{2}{3}J^{-5/3}\bar{\boldsymbol{\Sigma}}_H : \boldsymbol{H}\right)\boldsymbol{H}.\end{aligned} \tag{3.33}$$

Concerning (3.29), the stresses $\bar{\boldsymbol{\Sigma}}_F$, $\bar{\boldsymbol{\Sigma}}_H$ and p are given as

$$\bar{\boldsymbol{\Sigma}}_F = 2\alpha\bar{\boldsymbol{F}}, \qquad \bar{\boldsymbol{\Sigma}}_H = 3\beta(\bar{\boldsymbol{H}} : \bar{\boldsymbol{H}})^{1/2}\bar{\boldsymbol{H}} \qquad \text{and} \qquad p = \kappa(J-1) - d\gamma(\theta - \theta_{\text{ref}})\kappa, \tag{3.34}$$

and the entropy reads

$$\eta = c\ln\left(\frac{\theta}{\theta_{\text{ref}}}\right) + d\gamma\kappa(J-1). \tag{3.35}$$

For the Newton-Raphson iteration a linearization is required. Regarding (3.32), the linearization reads

$$\begin{aligned}\mathrm{D}^2\bar{\Psi}[\delta\boldsymbol{\varphi}, \delta\theta, \Delta\boldsymbol{\varphi}, \Delta\theta] &= \begin{bmatrix}\nabla(\delta\boldsymbol{\varphi}) : & \delta\theta\end{bmatrix}\begin{bmatrix}\mathrm{D}\boldsymbol{P}[\nabla(\Delta\boldsymbol{\varphi}), \Delta\theta] \\ -\mathrm{D}\eta[\nabla(\Delta\boldsymbol{\varphi}), \Delta\theta]\end{bmatrix} \\ &= \begin{bmatrix}\nabla(\delta\boldsymbol{\varphi}) : \\ \nabla(\delta\boldsymbol{\varphi}) \times \boldsymbol{F} : \\ \nabla(\delta\boldsymbol{\varphi}) : \boldsymbol{H} \\ \delta\theta\end{bmatrix}^{\mathrm{T}} [\mathbb{H}_{\bar{\Psi}}] \begin{bmatrix} : \nabla(\Delta\boldsymbol{\varphi}) \\ : \nabla(\Delta\boldsymbol{\varphi}) \times \boldsymbol{F} \\ \nabla(\Delta\boldsymbol{\varphi}) : \boldsymbol{H} \\ \Delta\theta\end{bmatrix} \\ &\quad + (\boldsymbol{\Sigma}_{\boldsymbol{H}} + \Sigma_J\boldsymbol{F}) : (\nabla(\delta\boldsymbol{\varphi}) \times \nabla(\Delta\boldsymbol{\varphi})),\end{aligned} \tag{3.36}$$

where the Hessian operator is defined as

$$[\mathbb{H}_{\bar{\Psi}}] = \begin{bmatrix}\frac{\partial^2\bar{\Psi}}{\partial\boldsymbol{F}\partial\boldsymbol{F}} & \frac{\partial^2\bar{\Psi}}{\partial\boldsymbol{F}\partial\boldsymbol{H}} & \frac{\partial^2\bar{\Psi}}{\partial\boldsymbol{F}\partial J} & \frac{\partial^2\bar{\Psi}}{\partial\boldsymbol{F}\partial\theta} \\ \frac{\partial^2\bar{\Psi}}{\partial\boldsymbol{H}\partial\boldsymbol{F}} & \frac{\partial^2\bar{\Psi}}{\partial\boldsymbol{H}\partial\boldsymbol{H}} & \frac{\partial^2\bar{\Psi}}{\partial\boldsymbol{H}\partial J} & \frac{\partial^2\bar{\Psi}}{\partial\boldsymbol{H}\partial\theta} \\ \frac{\partial^2\bar{\Psi}}{\partial J\partial\boldsymbol{F}} & \frac{\partial^2\bar{\Psi}}{\partial J\partial\boldsymbol{H}} & \frac{\partial^2\bar{\Psi}}{\partial J\partial J} & \frac{\partial^2\bar{\Psi}}{\partial J\partial\theta} \\ \frac{\partial^2\bar{\Psi}}{\partial\theta\partial\boldsymbol{F}} & \frac{\partial^2\bar{\Psi}}{\partial\theta\partial\boldsymbol{H}} & \frac{\partial^2\bar{\Psi}}{\partial\theta\partial J} & \frac{\partial^2\bar{\Psi}}{\partial\theta\partial\theta}\end{bmatrix}. \tag{3.37}$$

The entries of the Hessian operator read

$$\begin{aligned}\frac{\partial^2\bar{\Psi}}{\partial\boldsymbol{F}\partial\boldsymbol{F}} &= 2\alpha J^{-2/3}\mathbb{I}, \\ \frac{\partial^2\bar{\Psi}}{\partial\boldsymbol{F}\partial J} &= -\frac{4}{3}\alpha J^{-5/3}\boldsymbol{F}, \\ \frac{\partial^2\bar{\Psi}}{\partial\boldsymbol{H}\partial\boldsymbol{H}} &= 3\beta J^{-2}((\boldsymbol{H} : \boldsymbol{H})^{-1/2}\boldsymbol{H} \otimes \boldsymbol{H} + (\boldsymbol{H} : \boldsymbol{H})^{1/2}\mathbb{I}), \\ \frac{\partial^2\bar{\Psi}}{\partial\boldsymbol{H}\partial J} &= -6\beta J^{-3}(\boldsymbol{H} : \boldsymbol{H})^{1/2}\boldsymbol{H}, \\ \frac{\partial^2\bar{\Psi}}{\partial J\partial J} &= \frac{10}{9}\alpha J^{-8/3}\boldsymbol{F} : \boldsymbol{F} + 6\beta J^{-4}(\boldsymbol{H} : \boldsymbol{H})^{3/2} + \kappa, \\ \frac{\partial^2\bar{\Psi}}{\partial J\partial\theta} &= -d\gamma\kappa, \\ \frac{\partial^2\bar{\Psi}}{\partial\theta\partial\theta} &= -\frac{c}{\theta},\end{aligned} \tag{3.38}$$

where $\mathbb{I}$ denotes the fourth order identity tensor. Note that $[\mathbb{H}_{\bar{\Psi}}]$ can be written as a 20×20 second order tensor by applying the Voigt notation. Moreover, the first term in $(3.36)_2$ represents the material part of the linearization, whereas the geometrical part is given via the second term. As shown in Bonet et al. [10], geometrical effects like buckling are solely expressed by the latter term.

3.3.3 Duhamel's law of heat conduction

A suitable constitutive relation for the computation of the Piola-Kirchhoff heat flux vector $\boldsymbol{Q}$ reads

$$\boldsymbol{Q}(\boldsymbol{F}, \theta, \nabla\theta) = -\boldsymbol{F}^{-1}\boldsymbol{K}(\theta)\boldsymbol{F}^{-\mathrm{T}}\nabla(\theta), \tag{3.39}$$

known as Duhamel's law of heat conduction in material description. Here, $\boldsymbol{K}$ denotes the positive semi-definite thermal conductivity tensor, which can be simplified in case of isotropic material to $\boldsymbol{K}(\theta) = K(\theta)\boldsymbol{I} = K_{\mathrm{ref}}(1 - \omega(\theta - \theta_{\mathrm{ref}}))\boldsymbol{I}$ with $K(\theta) \geq 0$. Inserting into (3.39) yields

$$\begin{aligned}\boldsymbol{Q}(\boldsymbol{F}, \theta, \nabla\theta) &= -K(\theta)\boldsymbol{F}^{-1}\boldsymbol{F}^{-\mathrm{T}}\nabla(\theta)\\ &= -K_{\mathrm{ref}}(1 - \omega(\theta - \theta_{\mathrm{ref}}))\boldsymbol{F}^{-1}\boldsymbol{F}^{-\mathrm{T}}\nabla(\theta),\end{aligned} \tag{3.40}$$

where ω is a softening parameter relying on the observation that the thermal conductivity decreases with increasing temperature for a large number of materials, see Holzapfel & Simo [47] and the references therein. Note that (3.40) represents the Fourier's law of heat conduction which satisfies (3.17).

3.3.4 Complementary energy

Assuming that the relationship between the work conjugate variables $\{\bar{\boldsymbol{F}}, \bar{\boldsymbol{H}}, J\}$ and $\{\boldsymbol{\Sigma}_{\boldsymbol{F}}, \boldsymbol{\Sigma}_{\boldsymbol{H}}, p\}$ is invertible, we define a complementary energy density function by means of the Legendre transformation

$$\bar{\Upsilon}(\bar{\boldsymbol{\Sigma}}_F, \bar{\boldsymbol{\Sigma}}_H, p, \theta) = \sup_{\bar{\boldsymbol{F}}, \bar{\boldsymbol{H}}, J} \{\bar{\boldsymbol{\Sigma}}_F : \bar{\boldsymbol{F}} + \bar{\boldsymbol{\Sigma}}_H : \bar{\boldsymbol{H}} + pJ - \bar{\Psi}(\bar{\boldsymbol{F}}, \bar{\boldsymbol{H}}, J, \theta)\}. \tag{3.41}$$

Regarding (3.29), this leads to

$$\begin{aligned}\bar{\Upsilon} &= \bar{\boldsymbol{\Sigma}}_F : \bar{\boldsymbol{F}}(\bar{\boldsymbol{\Sigma}}_F) + \bar{\boldsymbol{\Sigma}}_H : \bar{\boldsymbol{H}}(\bar{\boldsymbol{\Sigma}}_H) + pJ(p, \theta) - \Psi(\bar{\boldsymbol{F}}(\bar{\boldsymbol{\Sigma}}_F), \bar{\boldsymbol{H}}(\bar{\boldsymbol{\Sigma}}_H), J(p, \theta), \theta)\\ &= \frac{1}{4\alpha}\bar{\boldsymbol{\Sigma}}_F : \bar{\boldsymbol{\Sigma}}_F + \alpha d + \frac{2}{3\sqrt{3\beta}}(\bar{\boldsymbol{\Sigma}}_H : \bar{\boldsymbol{\Sigma}}_H)^{3/4} + \beta d^{3/2} + \frac{1}{2\kappa}p^2 + p + d\gamma(\theta - \theta_{\mathrm{ref}})p\\ &\quad - c\left(\theta - \theta_{\mathrm{ref}} - \theta \ln\left(\frac{\theta}{\theta_{\mathrm{ref}}}\right)\right) + \frac{\kappa}{2}(d\gamma(\theta - \theta_{\mathrm{ref}}))^2,\end{aligned} \tag{3.42}$$

where expressions for $\bar{\boldsymbol{F}}(\bar{\boldsymbol{\Sigma}}_F)$, $\bar{\boldsymbol{H}}(\bar{\boldsymbol{\Sigma}}_H)$ and $J(p, \theta)$ may deduced directly by inverting the relations given in (3.34). Note that $\bar{\Upsilon}$ corresponds to the negative Gibbs energy density.

3.4 Variational formulation

Based on the derivations in section 3.2 we summarize the strong form of the thermomechanical coupled initial boundary value problem as follows

$$\begin{aligned} \rho_0 \dot{\boldsymbol{\varphi}} &= \boldsymbol{\pi} \\ \dot{\boldsymbol{\pi}} &= \operatorname{Div} \boldsymbol{P} + \boldsymbol{B} \quad , \\ \theta \dot{\eta} &= -\operatorname{Div} \boldsymbol{Q} + R \end{aligned} \tag{3.43}$$

where $\boldsymbol{\pi}$ denotes the linear momentum. Taking into account the constitutive relations discussed in section 3.3, the above set of equations has to be satisfied for all $\boldsymbol{X} \in \mathcal{B}_0$ and $t \in \mathcal{T}$. In order to complete the initial boundary value problem, the equations in (3.43) have to be supplemented by appropriate initial and boundary conditions. We specify initial conditions as

$$\boldsymbol{\varphi}(\boldsymbol{X}, 0) = \boldsymbol{X}, \qquad \dot{\boldsymbol{\varphi}}(\boldsymbol{X}, 0) = \boldsymbol{v}_0 \qquad \text{and} \qquad \theta(\boldsymbol{X}, 0) = \theta_0 \quad \text{in}\, \mathcal{B}_0, \tag{3.44}$$

where $\boldsymbol{v}_0$ and θ_0 are prescribed fields at $t = 0$, namely the initial velocity field and the initial temperature field. Moreover, the boundary conditions for the mechanical field are specified as follows

$$\boldsymbol{\varphi} = \bar{\boldsymbol{\varphi}} \quad \text{on}\, \partial\mathcal{B}_0^{\varphi} \times \mathcal{T} \qquad \text{and} \qquad \boldsymbol{T} = \bar{\boldsymbol{T}} \quad \text{on}\, \partial\mathcal{B}_0^{T} \times \mathcal{T}, \tag{3.45}$$

where $\partial\mathcal{B}_0^{\varphi}$ and $\partial\mathcal{B}_0^{T}$ are subsets of $\partial\mathcal{B}_0$ with properties $\partial\mathcal{B}_0^{\varphi} \cup \partial\mathcal{B}_0^{T} = \partial\mathcal{B}_0$ and $\partial\mathcal{B}_0^{\varphi} \cap \partial\mathcal{B}_0^{T} = \emptyset$. Similarly, the boundary conditions of the thermal field are given by

$$\theta = \bar{\theta} \quad \text{on}\, \partial\mathcal{B}_0^{\theta} \times \mathcal{T} \qquad \text{and} \qquad Q_{\mathrm{N}} = \bar{Q}_{\mathrm{N}} \quad \text{on}\, \partial\mathcal{B}_0^{Q} \times \mathcal{T}, \tag{3.46}$$

where $\partial\mathcal{B}_0^{\theta}$ and $\partial\mathcal{B}_0^{Q}$ are subsets of $\partial\mathcal{B}_0$ with properties $\partial\mathcal{B}_0^{\theta} \cup \partial\mathcal{B}_0^{Q} = \partial\mathcal{B}_0$ and $\partial\mathcal{B}_0^{\theta} \cap \partial\mathcal{B}_0^{Q} = \emptyset$. Furthermore, $\bar{\boldsymbol{\varphi}}$, $\bar{\boldsymbol{T}}$, $\bar{\theta}$ and $\bar{Q}_{\mathrm{N}}$ are prescribed fields.

3.4.1 Principle of virtual work

We define the space of the virtual variation of the geometry as

$$\mathcal{V}^{\varphi} = \{\delta\boldsymbol{\varphi} \in \mathcal{H}^1(\mathcal{B}_0) | \delta\boldsymbol{\varphi} = \boldsymbol{0}\, \text{on}\, \partial\mathcal{B}_0^{\varphi} \times \mathcal{T}\} \tag{3.47}$$

and the corresponding space of the virtual variation of the absolute temperature as

$$\mathcal{V}^{\theta} = \{\delta\theta \in \mathcal{H}^1(\mathcal{B}_0) | \delta\theta = 0\, \text{on}\, \partial\mathcal{B}_0^{\theta} \times \mathcal{T}\}. \tag{3.48}$$

Scalar multiplication of $(3.43)_2$ and $(3.43)_3$ by the virtual variations and subsequent integration lead to

$$\begin{aligned} \int_{\mathcal{B}_0} \delta\boldsymbol{\varphi} \cdot \dot{\boldsymbol{\pi}} \mathrm{d}V &= \int_{\mathcal{B}_0} \delta\boldsymbol{\varphi} \cdot (\operatorname{Div}(\boldsymbol{P}) + \boldsymbol{B}) \mathrm{d}V, \\ \int_{\mathcal{B}_0} \delta\theta\theta\dot{\eta} \mathrm{d}V &= \int_{\mathcal{B}_0} \delta\theta(R - \operatorname{Div}(\boldsymbol{Q})) \mathrm{d}V. \end{aligned} \tag{3.49}$$

Eventually, integration by parts along with the divergence theorem yields

$$
\begin{aligned}
G_\varphi &= \int\limits_{\mathcal{B}_0} (\delta\boldsymbol{\varphi}\cdot\dot{\boldsymbol{\pi}} + \boldsymbol{P} : \nabla(\delta\boldsymbol{\varphi}))\mathrm{d}V - \int\limits_{\mathcal{B}_0} \delta\boldsymbol{\varphi}\cdot\boldsymbol{B}\mathrm{d}V - \int\limits_{\partial\mathcal{B}_0^T} \delta\boldsymbol{\varphi}\cdot\bar{\boldsymbol{T}}\mathrm{d}A = 0,\\
G_\theta &= \int\limits_{\mathcal{B}_0} (\delta\theta\theta\dot{\eta} - \boldsymbol{Q}\cdot\nabla(\delta\theta))\mathrm{d}V - \int\limits_{\mathcal{B}_0} \delta\theta R\mathrm{d}V - \int\limits_{\partial\mathcal{B}_0^Q} \delta\theta\bar{Q}_{\mathrm{N}}\mathrm{d}A = 0.
\end{aligned}
\tag{3.50}
$$

These equations have to hold for all $\delta\boldsymbol{\varphi} \in \mathcal{V}^\varphi$ and $\delta\theta \in \mathcal{V}^\theta$. While we state the balance of linear momentum and the balance of energy in weak form, we retain the kinematic relation $(3.43)_1$ as well as the constitutive relations (3.23) and (3.24) in local form. Note also that the first Piola-Kirchhoff stress tensor $\boldsymbol{P}$ can be calculated via (3.33) depending on the choice of the free Helmholtz energy density function.

3.4.2 Hellinger-Reissner type variational principle

In the sense of a Hellinger-Reissner type variational principle, the original set of primal variables is extended by stresses. Accordingly, we define $\{\boldsymbol{\varphi}, \theta, \bar{\boldsymbol{\Sigma}}_F^*, \bar{\boldsymbol{\Sigma}}_H^*, p^*\}$ as new set of primal variables, where the star indicates the stresses as primal variables. Next, we perform two consecutive Legendre transformations[2]

$$
\begin{aligned}
e(\boldsymbol{F},\eta) &= \bar{\Psi}(\bar{\boldsymbol{F}}, \bar{\boldsymbol{H}}, J, \theta) + \theta\eta\\
&= -\bar{\Upsilon}^*(\bar{\boldsymbol{\Sigma}}_F^*, \bar{\boldsymbol{\Sigma}}_H^*, p^*, \theta) + \bar{\boldsymbol{\Sigma}}_F^* : \bar{\boldsymbol{F}} + \bar{\boldsymbol{\Sigma}}_H^* : \bar{\boldsymbol{H}} + p^* J + \theta\eta^*
\end{aligned}
\tag{3.51}
$$

and recast (3.10) in terms of the complementary energy density function as follows

$$
\begin{aligned}
\int\limits_{\mathcal{B}_0} \Bigg(&(\dot{\boldsymbol{\pi}} - \mathrm{Div}(\boldsymbol{P}^*) - \boldsymbol{B})\cdot\dot{\boldsymbol{\varphi}} + \left(\bar{\boldsymbol{F}} - \frac{\partial\bar{\Upsilon}^*}{\partial\bar{\boldsymbol{\Sigma}}_F^*}\right) : \dot{\bar{\boldsymbol{\Sigma}}}_F^* + \left(\bar{\boldsymbol{H}} - \frac{\partial\bar{\Upsilon}^*}{\partial\bar{\boldsymbol{\Sigma}}_H^*}\right) : \dot{\bar{\boldsymbol{\Sigma}}}_H^*\\
&+ \left(J - \frac{\partial\bar{\Upsilon}^*}{\partial p^*}\right)\dot{p}^* + \left(\eta^* - \frac{\partial\bar{\Upsilon}^*}{\partial\theta}\right)\dot{\theta} + (\theta\dot{\eta}^* + \mathrm{Div}(\boldsymbol{Q}) - R)\Bigg)\mathrm{d}V = 0.
\end{aligned}
\tag{3.52}
$$

Assuming that the rates of change occurring in (3.52) can be chosen arbitrary, we apply suitable substitutions, namely $\dot{\boldsymbol{\varphi}} = \delta\boldsymbol{\varphi}$, $\dot{\bar{\boldsymbol{\Sigma}}}_F^* = \delta\bar{\boldsymbol{\Sigma}}_F$, $\dot{\bar{\boldsymbol{\Sigma}}}_H^* = \delta\bar{\boldsymbol{\Sigma}}_H$ and $\dot{p}^* = \delta p$. Together with the virtual variation of the temperature $\delta\theta$ applied to the latter terms in (3.52), this leads to

$$
\begin{aligned}
G_\varphi^{\mathrm{hr}} &= \int\limits_{\mathcal{B}_0} (\delta\boldsymbol{\varphi}\cdot\dot{\boldsymbol{\pi}} + \boldsymbol{P}^* : \nabla(\delta\boldsymbol{\varphi}))\mathrm{d}V - \int\limits_{\mathcal{B}_0} \delta\boldsymbol{\varphi}\cdot\boldsymbol{B}\mathrm{d}V - \int\limits_{\partial\mathcal{B}_0^T} \delta\boldsymbol{\varphi}\cdot\bar{\boldsymbol{T}}\mathrm{d}A = 0,\\
G_\theta^{\mathrm{hr}} &= \int\limits_{\mathcal{B}_0} (\delta\theta\theta\dot{\eta}^* - \boldsymbol{Q}\cdot\nabla(\delta\theta))\mathrm{d}V - \int\limits_{\mathcal{B}_0} \delta\theta R\mathrm{d}V - \int\limits_{\partial\mathcal{B}_0^Q} \delta\theta\bar{Q}_{\mathrm{N}}\mathrm{d}A = 0,
\end{aligned}
\tag{3.53}
$$

[2] Physical quantities which are subsequently considered as functions of $\bar{\boldsymbol{\Sigma}}_F^*$,$\bar{\boldsymbol{\Sigma}}_H^*$ and p^* are also indicated by a star.

supplemented by the geometric compatibility conditions

$$G^{\mathrm{hr}}_{\Sigma_F} = \int_{\mathcal{B}_0} \left(\bar{\boldsymbol{F}} - \frac{\partial \bar{\Upsilon}^*}{\partial \bar{\boldsymbol{\Sigma}}^*_F} \right) : \delta \bar{\boldsymbol{\Sigma}}_F \mathrm{d}V = 0,$$
$$G^{\mathrm{hr}}_{\Sigma_H} = \int_{\mathcal{B}_0} \left(\bar{\boldsymbol{H}} - \frac{\partial \bar{\Upsilon}^*}{\partial \bar{\boldsymbol{\Sigma}}^*_H} \right) : \delta \bar{\boldsymbol{\Sigma}}_H \mathrm{d}V = 0, \tag{3.54}$$
$$G^{\mathrm{hr}}_{p} = \int_{\mathcal{B}_0} \left(J - \frac{\partial \bar{\Upsilon}^*}{\partial p^*} \right) \delta p \mathrm{d}V = 0.$$

Note that the thermal compatibility condition $\eta^* = \partial \bar{\Upsilon}^* / \partial \theta$ is satisfied locally, but can also be enforced in variational sense by introducing the entropy as primal variable, similar to (3.54). In addition, we define the first Piola-Kirchhoff stress tensor $\boldsymbol{P}^*$ as

$$\boldsymbol{P}^* = J^{-1/3} \bar{\boldsymbol{\Sigma}}^*_F + J^{-2/3} \bar{\boldsymbol{\Sigma}}^*_H \times \boldsymbol{F} + \left(p^* - \frac{1}{3} J^{-4/3} \bar{\boldsymbol{\Sigma}}^*_F : \boldsymbol{F} - \frac{2}{3} J^{-5/3} \bar{\boldsymbol{\Sigma}}^*_H : \boldsymbol{H} \right) \boldsymbol{H}. \tag{3.55}$$

Obviously, (3.53) merely differs from (3.50) by the definition of the first Piola-Kirchhoff stress tensor and the entropy.

3.5 Discrete setting

In this section we perform the discretization in space and time for both, the standard displacement-temperature based approach as well as the mixed approach based on the Hellinger-Reissner type variational principle. For the latter approach, we apply a novel discretization scheme which relies on developments suggested in Bonet et al. [10]. In particular, we utilize B-spline based shape functions for the spatial discretization of the geometry and the temperature, whereas a discontinuous interpolation across element borders is applied to stress components which are primal variables within the Hellinger-Reissner type variational principle. This allows us for the reduction of the global system by applying a standard condensation procedure related to the stress unknowns. Moreover, an implicit time integration scheme is applied to the semi-discrete coupled thermoelastic problem. In particular, we apply a second order accurate mid-point evaluation to obtain a set of non-linear algebraic equations to be solved via the Newton-Raphson iteration scheme.

3.5.1 Spatial discretization – Displacement-temperature based approach

The finite element approximations of the geometry $\boldsymbol{\varphi}$ and its variation $\delta\boldsymbol{\varphi}$ are given as

$$\boldsymbol{\varphi}^{\mathrm{h}} = \sum_{A \in \mathcal{I}} R^A \boldsymbol{q}_A \qquad \text{and} \qquad \delta\boldsymbol{\varphi}^{\mathrm{h}} = \sum_{A \in \mathcal{I}} R^A \delta \boldsymbol{q}_A \tag{3.56}$$

and the corresponding approximations of the absolute temperature θ and its variation $\delta\theta$ read

$$\theta^{\mathrm{h}} = \sum_{A\in\mathcal{I}} R^A \Theta_A \qquad \text{and} \qquad \delta\theta^{\mathrm{h}} = \sum_{A\in\mathcal{I}} R^A \delta\Theta_A. \tag{3.57}$$

Here, $\boldsymbol{q}_A$ denotes the net of control points related to the current geometry and Θ_A are the control variables for the temperature field. For both we use the same B-spline based approximations, where $R^A(\boldsymbol{X})$ are global basis functions and $\mathcal{I}$ is the index set of active basis functions over all existent hierarchical levels.

In order to derive the semi-discrete counterpart of $(3.50)_1$, we insert (3.56) into the corresponding terms of the mechanical balance equation. Beginning with the first term, we obtain

$$\begin{aligned}\int_{\mathcal{B}_0} \delta\boldsymbol{\varphi}^{\mathrm{h}} \dot{\boldsymbol{\pi}}^{\mathrm{h}} \mathrm{d}V &= \int_{\mathcal{B}_0} \delta\boldsymbol{\varphi}^{\mathrm{h}} \rho_0 \ddot{\boldsymbol{\varphi}}^{\mathrm{h}} \mathrm{d}V \\ &= \delta\boldsymbol{q}_A \cdot M^{AB} \dot{\boldsymbol{v}}_B,\end{aligned} \tag{3.58}$$

where the coefficients of the consistent mass matrix are given as

$$M^{AB} = \int_{\mathcal{B}_0} \rho_0 R^A R^B \mathrm{d}V \tag{3.59}$$

and $\boldsymbol{v}_A = \dot{\boldsymbol{q}}_A$ denotes the control variables for the velocity field. Moreover, the semi-discrete virtual work of internal forces reads

$$\int_{\mathcal{B}_0} \boldsymbol{P}^{\mathrm{h}} : \nabla(\delta\boldsymbol{\varphi}^{\mathrm{h}}) \mathrm{d}V = \delta\boldsymbol{q}_A \cdot \int_{\mathcal{B}_0} \boldsymbol{P}^{\mathrm{h}} \nabla(R^A) \mathrm{d}V. \tag{3.60}$$

Recalling $(3.33)_2$, we obtain the discrete first Piola-Kirchhoff stress tensor via

$$\begin{aligned}\boldsymbol{P}^{\mathrm{h}} &= J^{\mathrm{h}^{-1/3}} \bar{\boldsymbol{\Sigma}}_F^{\mathrm{h}} + J^{\mathrm{h}^{-2/3}} \bar{\boldsymbol{\Sigma}}_H^{\mathrm{h}} \times \boldsymbol{F}^{\mathrm{h}} \\ &+ \left(p^{\mathrm{h}} - \frac{1}{3} J^{\mathrm{h}^{-4/3}} \bar{\boldsymbol{\Sigma}}_F^{\mathrm{h}} : \boldsymbol{F}^{\mathrm{h}} - \frac{2}{3} J^{\mathrm{h}^{-5/3}} \bar{\boldsymbol{\Sigma}}_H^{\mathrm{h}} : \boldsymbol{H}^{\mathrm{h}} \right) \boldsymbol{H}^{\mathrm{h}},\end{aligned} \tag{3.61}$$

where

$$\boldsymbol{F}^{\mathrm{h}} = \frac{\partial \boldsymbol{\varphi}^{\mathrm{h}}}{\partial \boldsymbol{X}} = \boldsymbol{q}_A \otimes \nabla(R^A) \tag{3.62}$$

is the discrete version of the deformation gradient. The strain measures $\boldsymbol{H}^{\mathrm{h}}$, J^{h}, $\bar{\boldsymbol{F}}^{\mathrm{h}}$ and $\bar{\boldsymbol{H}}^{\mathrm{h}}$ follow in a straightforward manner by using the introduced definitions and the discrete stress components can be achieved via $\bar{\Psi}^{\mathrm{h}} = \bar{\Psi}^{\mathrm{h}}(\bar{\boldsymbol{F}}^{\mathrm{h}}, \bar{\boldsymbol{H}}^{\mathrm{h}}, J^{\mathrm{h}}, \theta^{\mathrm{h}})$ as follows

$$\bar{\boldsymbol{\Sigma}}_F^{\mathrm{h}} = \frac{\partial \bar{\Psi}^{\mathrm{iso,h}}}{\partial \bar{\boldsymbol{F}}^{\mathrm{h}}}, \qquad \bar{\boldsymbol{\Sigma}}_H^{\mathrm{h}} = \frac{\partial \bar{\Psi}^{\mathrm{iso,h}}}{\partial \bar{\boldsymbol{H}}^{\mathrm{h}}} \qquad \text{and} \qquad p^{\mathrm{h}} = \frac{\partial(\bar{\Psi}^{\mathrm{vol,h}} + \bar{\Psi}^{\mathrm{cpl,h}})}{\partial J^{\mathrm{h}}}. \tag{3.63}$$

Eventually, the term of virtual work of external forces in semi-discrete setting reads

$$\begin{aligned}\int_{\mathcal{B}_0} \delta\boldsymbol{\varphi}^{\mathrm{h}} \cdot \boldsymbol{B} \mathrm{d}V + \int_{\partial\mathcal{B}_0^T} \delta\boldsymbol{\varphi}^{\mathrm{h}} \cdot \bar{\boldsymbol{T}} \mathrm{d}A &= \delta\boldsymbol{q}_A \cdot \left[\int_{\mathcal{B}_0} R^A \boldsymbol{B} \mathrm{d}V + \int_{\partial\mathcal{B}_0^T} R^A \bar{\boldsymbol{T}} \mathrm{d}A \right] \\ &= \delta\boldsymbol{q}_A \cdot \boldsymbol{F}^{\mathrm{ext},A}\end{aligned} \tag{3.64}$$

and thus the semi-discrete counterpart of $(3.50)_1$ can be written as

$$\begin{aligned} G_{\varphi}^{\mathrm{h}} &= \delta \boldsymbol{q}_A \cdot \left[M^{AB} \dot{\boldsymbol{v}}_B + \int\limits_{\mathcal{B}_0} \boldsymbol{P}^{\mathrm{h}} \nabla(R^A) \mathrm{d}V - \boldsymbol{F}^{\mathrm{ext},A} \right] \\ &= \delta \boldsymbol{q}_A \cdot \boldsymbol{R}_{\varphi}^A . \end{aligned} \tag{3.65}$$

Next, we insert (3.57) into the terms of the weak form of thermal balance equation given by $(3.50)_2$. The first term in semi-discrete setting reads

$$\begin{aligned} \int\limits_{\mathcal{B}_0} \delta\theta^{\mathrm{h}} \theta^{\mathrm{h}} \dot{\eta}^{\mathrm{h}} &= \delta\Theta_A \int\limits_{\mathcal{B}_0} R^A \dot{\eta}^{\mathrm{h}} \theta^{\mathrm{h}} \mathrm{d}V \\ &= \delta\Theta_A \int\limits_{\mathcal{B}_0} R^A \Gamma^{\mathrm{h}} \mathrm{d}V, \end{aligned} \tag{3.66}$$

where the semi-discrete entropy is obtained via

$$\eta^{\mathrm{h}} = -\frac{\partial(\bar{\Psi}^{\mathrm{the,h}} + \bar{\Psi}^{\mathrm{cpl,h}})}{\partial\theta^{\mathrm{h}}}. \tag{3.67}$$

The semi-discrete version of the second term reads

$$\int\limits_{\mathcal{B}_0} \boldsymbol{Q}^{\mathrm{h}} \cdot \nabla(\delta\theta^{\mathrm{h}}) \mathrm{d}V = \delta\Theta_A \int\limits_{\mathcal{B}_0} \boldsymbol{Q}^{\mathrm{h}} \cdot \nabla(R^A) \mathrm{d}V, \tag{3.68}$$

with the discrete Piola-Kirchhoff heat flux vector

$$\boldsymbol{Q}^{\mathrm{h}} = -K_{\mathrm{ref}}(1 - \omega(\theta^{\mathrm{h}} - \theta_{\mathrm{ref}})) {\boldsymbol{F}^{\mathrm{h}}}^{-1} {\boldsymbol{F}^{\mathrm{h}}}^{-T} \nabla(\theta^{\mathrm{h}}). \tag{3.69}$$

Finally, the terms of external contributions in semi-discrete setting are given as

$$\begin{aligned} \int\limits_{\mathcal{B}_0} \delta\theta^{\mathrm{h}} R \mathrm{d}V + \int\limits_{\partial\mathcal{B}_0^Q} \delta\theta^{\mathrm{h}} \bar{Q}_{\mathrm{N}} \mathrm{d}A &= \delta\Theta_A \left[\int\limits_{\mathcal{B}_0} R^A R \mathrm{d}V + \int\limits_{\partial\mathcal{B}_0^Q} R^A \bar{Q}_{\mathrm{N}} \mathrm{d}A \right] \\ &= \delta\Theta_A \bar{Q}^{\mathrm{ext},A} \end{aligned} \tag{3.70}$$

and thus the semi-discrete counterpart of $(3.50)_2$ can be written as

$$\begin{aligned} G_{\theta}^{\mathrm{h}} &= \delta\Theta_A \left[\int\limits_{\mathcal{B}_0} R^A \Gamma^{\mathrm{h}} \mathrm{d}V - \int\limits_{\mathcal{B}_0} \boldsymbol{Q}^{\mathrm{h}} \cdot \nabla(R^A) \mathrm{d}V - \bar{Q}^A \right] \\ &= \delta\Theta_A R_{\theta}^A . \end{aligned} \tag{3.71}$$

3.5.2 Spatial discretization – Mixed approach

Alternatively, employing the proposed Hellinger-Reissner type variational principle, the finite element approximations of the stress components read

$$\bar{\boldsymbol{\Sigma}}_F^{*,\mathrm{h}} = \sum_{A=1}^{n_{\Sigma_F}} N_{\Sigma_F}^A \bar{\boldsymbol{\Sigma}}_{F,A}, \qquad \bar{\boldsymbol{\Sigma}}_H^{*,\mathrm{h}} = \sum_{A=1}^{n_{\Sigma_H}} N_{\Sigma_H}^A \bar{\boldsymbol{\Sigma}}_{H,A} \qquad \text{and} \qquad p^{*,\mathrm{h}} = \sum_{A=1}^{n_p} N_p^A p_A, \tag{3.72}$$

where $N_{\Sigma_F}^A$ and $N_{\Sigma_H}^A$ are standard linear interpolation functions and $N_{\Sigma_J}^A$ denotes a constant interpolation. Using these approximations, the semi-discrete first Piola-Kirchhoff stress tensor $\boldsymbol{P}^{*,\mathrm{h}}$ is given in a straightforward manner, whereas the discrete entropy is determined from $\bar{\Upsilon}^{*,\mathrm{h}} = \bar{\Upsilon}^{*,\mathrm{h}}(\bar{\boldsymbol{\Sigma}}_F^{*,\mathrm{h}}, \bar{\boldsymbol{\Sigma}}_H^{*,\mathrm{h}}, p^{*,\mathrm{h}}, \theta^{\mathrm{h}})$ as follows

$$\eta^{*,\mathrm{h}} = \frac{\partial \bar{\Upsilon}^{*,\mathrm{h}}}{\partial \theta^{\mathrm{h}}}. \tag{3.73}$$

The semi-discrete settings $G_\varphi^{\mathrm{hr,h}}$ and $G_\theta^{\mathrm{hr,h}}$ are then analogous to (3.65) and (3.71), respectively. Next, we define the approximations of the virtual variations of the stress components via

$$\delta\bar{\boldsymbol{\Sigma}}_F^{\mathrm{h}} = \sum_{A=1}^{n_{\Sigma_F}} N_{\Sigma_F}^A \delta\bar{\boldsymbol{\Sigma}}_{F,A}, \qquad \delta\bar{\boldsymbol{\Sigma}}_H^{\mathrm{h}} = \sum_{A=1}^{n_{\Sigma_H}} N_{\Sigma_H}^A \delta\bar{\boldsymbol{\Sigma}}_{H,A} \qquad \text{and} \qquad \delta p^{\mathrm{h}} = \sum_{A=1}^{n_p} N_p^A \delta p_A \tag{3.74}$$

such that the geometric compatibility conditions in semi-discrete setting read

$$\begin{aligned}
G_{\Sigma_F}^{\mathrm{hr,h}} &= \delta\bar{\boldsymbol{\Sigma}}_{F,A} \int_{\mathcal{B}_0} \left(\bar{\boldsymbol{F}}^{\mathrm{h}} - \frac{\partial \bar{\Upsilon}^{*,\mathrm{h}}}{\partial \bar{\boldsymbol{\Sigma}}_F^{*,\mathrm{h}}} \right) N_{\Sigma_F}^A \mathrm{d}V \\
&= \delta\bar{\boldsymbol{\Sigma}}_{F,A} : \boldsymbol{R}_{\Sigma_F}^{\mathrm{hr},A}, \\
G_{\Sigma_H}^{\mathrm{hr,h}} &= \delta\bar{\boldsymbol{\Sigma}}_{H,A} : \int_{\mathcal{B}_0} \left(\bar{\boldsymbol{H}}^{\mathrm{h}} - \frac{\partial \bar{\Upsilon}^{*,\mathrm{h}}}{\partial \bar{\boldsymbol{\Sigma}}_H^{*,\mathrm{h}}} \right) N_{\Sigma_H}^A \mathrm{d}V \\
&= \delta\bar{\boldsymbol{\Sigma}}_{H,A} : \boldsymbol{R}_{\Sigma_H}^{\mathrm{hr},A}, \\
G_p^{\mathrm{hr,h}} &= \delta p_A \int_{\mathcal{B}_0} \left(J^{\mathrm{h}} - \frac{\partial \bar{\Upsilon}^{*,\mathrm{h}}}{\partial p^{*,\mathrm{h}}} \right) N_p^A \mathrm{d}V \\
&= \delta p_A R_p^{\mathrm{hr},A}.
\end{aligned} \tag{3.75}$$

While we use a B-spline based approximation of arbitrary order for the displacement-temperature based approach, we discretize the various fields related to the mixed approach independently. In particular, the geometry as well as the temperature field are approximated by quadratic B-spline based shape functions, whereas linear interpolations for the stresses conjugate to the isochoric part of the deformation gradient and its cofactor and a constant interpolation for the stress conjugate to the Jacobian determinant are applied. Note that these linear and constant interpolations are given element by element, i.e. discontinuous across element borders to ensure a block diagonal stiffness matrix of the geometrical compatibility conditions. This enables us to apply a condensation procedure, see section 3.5.4 for details. Moreover, we obtain $n_{\Sigma_F} = n_{\Sigma_H} = 8E$ and $n_p = E$, where E is the total number of elements.

3.5.3 Temporal discretization – Displacement-temperature based approach

Next, we perform the discretization in time for the coupled thermoelastic problem under consideration. We subdivide the considered time interval $\mathcal{T}$ into a sequence of times $t_0, \ldots, t_n, t_{n+1}, \ldots, T$, where $(\bullet)_n$ and $(\bullet)_{n+1}$ denote the value of a given physical quantity at time t_n and t_{n+1}, respectively. Assume the state variables at t_n, given by $\boldsymbol{\varphi}_n^h$ and θ_n^h, are known and the time step size $\Delta t = t_{n+1} - t_n$ is given. Then we approximate the state at t_{n+1} using a second order accurate mid-point evaluation. More advanced energy consistent time integration schemes for non-linear thermoelasticity can be found in Hesch & Betsch [38]. The full-discrete setting of $(3.50)_1$ takes the form

$$\begin{aligned} G^{\mathrm{h}}_{\varphi,n,n+1} &= \delta \boldsymbol{q}_A \cdot \left[M^{AB} \frac{\boldsymbol{v}_{B,n+1} - \boldsymbol{v}_{B,n}}{\Delta t} + \int_{\mathcal{B}_0} \nabla(R^A) \cdot \boldsymbol{P}^{\mathrm{h}}_{n,n+1} \mathrm{d}V - \boldsymbol{F}^{\mathrm{ext},A}_{n+1/2} \right] \\ &= \delta \boldsymbol{q}_A \cdot \boldsymbol{R}^A_{\varphi,n,n+1} \end{aligned} \tag{3.76}$$

and the full-discrete setting of $(3.50)_2$ reads

$$\begin{aligned} G^{\mathrm{h}}_{\theta,n,n+1} &= \delta \Theta_A \left[\int_{\mathcal{B}_0} \Gamma^{\mathrm{h}}_{n,n+1} R^A \mathrm{d}V - \int_{\mathcal{B}_0} \boldsymbol{Q}^{\mathrm{h}}_{n,n+1} \cdot \nabla(R^A) \mathrm{d}V - \bar{Q}^A_{n+1/2} \right] \\ &= \delta \Theta_A R^A_{\theta,n,n+1} \,. \end{aligned} \tag{3.77}$$

The above full-discrete weak formulations are supplemented by the mid-point type approximation of the kinematic relationship given as

$$\frac{\boldsymbol{q}_{A,n+1} - \boldsymbol{q}_{A,n}}{\Delta t} = \boldsymbol{v}_{A,n+1/2}, \tag{3.78}$$

where $\boldsymbol{v}_{A,n+1/2} = (\boldsymbol{v}_{A,n+1} + \boldsymbol{v}_{A,n})/2$. In the above formulas, $(\bullet)_{n,n+1}$ denotes the value of a given physical quantity evaluated via the mid-point configuration of the primal variables

$$\boldsymbol{\varphi}^{\mathrm{h}}_{n+1/2} = \frac{\boldsymbol{\varphi}^{\mathrm{h}}_{n+1} + \boldsymbol{\varphi}^{\mathrm{h}}_{n}}{2} \quad \text{and} \quad \theta^{\mathrm{h}}_{n+1/2} = \frac{\theta^{\mathrm{h}}_{n+1} + \theta^{\mathrm{h}}_{n}}{2}. \tag{3.79}$$

Moreover, we apply the central difference quotient to the entropy rate $\dot{\eta}^{\mathrm{h}}$ such that we obtain

$$\Gamma^{\mathrm{h}}_{n,n+1} = \frac{\eta^{\mathrm{h}}_{n+1} - \eta^{\mathrm{h}}_{n}}{\Delta t} \theta^{\mathrm{h}}_{n+1/2}. \tag{3.80}$$

Linearization

The global system to be solved within the Newton-Raphson iteration reads [3]

$$\begin{bmatrix} \boldsymbol{K}_{\varphi\varphi} & \boldsymbol{K}_{\varphi\theta} \\ \boldsymbol{K}_{\theta\varphi} & \boldsymbol{K}_{\theta\theta} \end{bmatrix} \begin{bmatrix} \Delta\boldsymbol{q} \\ \Delta\boldsymbol{\Theta} \end{bmatrix} = \begin{bmatrix} \boldsymbol{R}_{\varphi} \\ \boldsymbol{R}_{\theta} \end{bmatrix}, \tag{3.81}$$

where $\boldsymbol{R}_{\varphi}$ and $\boldsymbol{R}_{\theta}$ are given by (3.76) and (3.77), respectively. Furthermore, $\Delta\boldsymbol{q} = [\Delta\boldsymbol{q}_1, \dots, \Delta\boldsymbol{q}_{|\mathcal{I}|}]$ and $\Delta\boldsymbol{\Theta} = [\Delta\Theta_1, \dots, \Delta\Theta_{|\mathcal{I}|}]$ represent the corresponding incremental variation of the control variables for the geometry and the temperature at time t_{n+1}. Note that $\boldsymbol{K}_{\varphi\theta} \neq (\boldsymbol{K}_{\theta\varphi})^{\mathrm{T}}$, which results in a non-symmetric system which is inefficient to solve and is associated with high numerical effort. An alternative solution technique is known as staggered method, where the monolithic system is splitted into smaller symmetric subsystems to be solved sequentially, see, e.g. Argyris & Doltsinis [4], Armero and Simo [6] and Suwannachit [88] for details.

3.5.4 Temporal discretization – Mixed approach

Concerning the Hellinger-Reissner type variational principle we define the mid-point configuration of the stresses as follows

$$\bar{\boldsymbol{\Sigma}}^{*,\mathrm{h}}_{F,n+1/2} = \frac{\bar{\boldsymbol{\Sigma}}^{*,\mathrm{h}}_{F,n+1} + \bar{\boldsymbol{\Sigma}}^{*,\mathrm{h}}_{F,n}}{2}, \quad \bar{\boldsymbol{\Sigma}}^{*,\mathrm{h}}_{H,n+1/2} = \frac{\bar{\boldsymbol{\Sigma}}^{*,\mathrm{h}}_{H,n+1} + \bar{\boldsymbol{\Sigma}}^{*,\mathrm{h}}_{H,n}}{2} \quad \text{and} \quad p^{*,\mathrm{h}}_{n+1/2} = \frac{p^{*,\mathrm{h}}_{n+1} + p^{*,\mathrm{h}}_{n}}{2}. \tag{3.82}$$

Note that we obtain $\bar{\boldsymbol{\Sigma}}^{\mathrm{h}}_{F,n,n+1}$, $\bar{\boldsymbol{\Sigma}}^{\mathrm{h}}_{H,n,n+1}$ and $p^{\mathrm{h}}_{n,n+1}$ via the full-discrete free Helmholtz energy density function $\Psi^{\mathrm{h}}_{n,n+1}$ within the displacement-temperature based approach, whereas we utilize the mid-point configuration of the stress components given by (3.82) to define the full-discrete version of first Piola-Kirchhoff stress tensor $\boldsymbol{P}^{*,\mathrm{h}}_{n,n+1}$ for the mixed approach. Additionally, the full-discrete setting of the geometrical compatibility conditions evaluated at the midpoint configuration reads

$$\begin{aligned} G^{\mathrm{hr,h}}_{\Sigma_F,n,n+1} &= \delta\bar{\boldsymbol{\Sigma}}_{F,A} : \int_{\mathcal{B}_0} \left(\bar{\boldsymbol{F}}^{\mathrm{h}}_{n,n+1} - \frac{\partial \bar{\Upsilon}^{\mathrm{h}}_{n,n+1}}{\partial \bar{\boldsymbol{\Sigma}}^{*,\mathrm{h}}_{F,n+1/2}} \right) N^A_{\Sigma_F} \mathrm{d}V \\ &= \delta\bar{\boldsymbol{\Sigma}}_{F,A} : \boldsymbol{R}^{\mathrm{hr},A}_{\Sigma_F,n,n+1}, \\ G^{\mathrm{hr,h}}_{\Sigma_H,n,n+1} &= \delta\bar{\boldsymbol{\Sigma}}_{H,A} : \int_{\mathcal{B}_0} \left(\bar{\boldsymbol{H}}^{\mathrm{h}}_{n,n+1} - \frac{\partial \bar{\Upsilon}^{\mathrm{h}}_{n,n+1}}{\partial \bar{\boldsymbol{\Sigma}}^{*,\mathrm{h}}_{H,n+1/2}} \right) N^A_{\Sigma_H} \mathrm{d}V \\ &= \delta\bar{\boldsymbol{\Sigma}}_{H,A} : \boldsymbol{R}^{\mathrm{hr},A}_{\Sigma_H,n,n+1}, \\ G^{\mathrm{hr,h}}_{p,n,n+1} &= \delta p_A \int_{\mathcal{B}_0} \left(J^{\mathrm{h}}_{n,n+1} - \frac{\partial \bar{\Upsilon}^{\mathrm{h}}_{n,n+1}}{\partial p^{*,\mathrm{h}}_{n+1/2}} \right) N^A_{p} \mathrm{d}V \\ &= \delta p_A R^{\mathrm{hr},A}_{p,n,n+1}. \end{aligned} \tag{3.83}$$

[3] Within the representation of the global system we neglect the specification of the time step and the iteration step for readability.

Linearization and condensation

Similar to (3.81), the global system related to the Hellinger-Reissner type variational principle reads

$$\begin{bmatrix} \boldsymbol{K}_{\varphi\varphi} & \boldsymbol{K}_{\varphi\theta} & \boldsymbol{K}_{\varphi\Sigma} \\ \boldsymbol{K}_{\theta\varphi} & \boldsymbol{K}_{\theta\theta} & \boldsymbol{K}_{\theta\Sigma} \\ \boldsymbol{K}_{\Sigma\varphi} & \boldsymbol{K}_{\Sigma\theta} & \boldsymbol{K}_{\Sigma\Sigma} \end{bmatrix}^{\text{hr}} \begin{bmatrix} \Delta\boldsymbol{q} \\ \Delta\boldsymbol{\Theta} \\ \Delta\boldsymbol{\Sigma} \end{bmatrix} = \begin{bmatrix} \boldsymbol{R}_{\varphi} \\ \boldsymbol{R}_{\theta} \\ \boldsymbol{R}_{\Sigma} \end{bmatrix}^{\text{hr}}, \tag{3.84}$$

where $\boldsymbol{R}_{\varphi}^{\text{hr}}$ and $\boldsymbol{R}_{\theta}^{\text{hr}}$ are analogous to $\boldsymbol{R}_{\varphi}$ and $\boldsymbol{R}_{\theta}$, respectively. In addition, $\boldsymbol{R}_{\Sigma}^{\text{hr}}$ is given via (3.83), whereas the associated incremental variations of the control variables of the stresses are summarized as $\Delta\boldsymbol{\Sigma} = [\Delta\bar{\boldsymbol{\Sigma}}_{F,1}, \dots, \Delta\bar{\boldsymbol{\Sigma}}_{F,n_{\Sigma_F}}, \Delta\bar{\boldsymbol{\Sigma}}_{H,1}, \dots, \Delta\bar{\boldsymbol{\Sigma}}_{H,n_{\Sigma_H}}, \Delta p_1, \dots, \Delta p_{n_p}]$.

Solving (3.84) with respect to $\Delta\boldsymbol{\Sigma}$ yields

$$\Delta\boldsymbol{\Sigma} = {\boldsymbol{K}_{\Sigma\Sigma}^{\text{hr}}}^{-1}(\boldsymbol{R}_{\Sigma}^{\text{hr}} - \boldsymbol{K}_{\Sigma\varphi}^{\text{hr}}\Delta\boldsymbol{q} - \boldsymbol{K}_{\Sigma\theta}^{\text{hr}}\Delta\boldsymbol{\Theta}). \tag{3.85}$$

Insertion into (3.84) gives us the condensed system

$$\begin{bmatrix} \boldsymbol{K}_{\varphi\varphi} - \boldsymbol{K}_{\varphi\Sigma}\boldsymbol{K}_{\Sigma\Sigma}^{-1}\boldsymbol{K}_{\Sigma\varphi} & -\boldsymbol{K}_{\varphi\Sigma}\boldsymbol{K}_{\Sigma\Sigma}^{-1}\boldsymbol{K}_{\Sigma\theta} \\ -\boldsymbol{K}_{\theta\Sigma}\boldsymbol{K}_{\Sigma\Sigma}^{-1}\boldsymbol{K}_{\Sigma\varphi} & \boldsymbol{K}_{\theta\theta} - \boldsymbol{K}_{\theta\Sigma}\boldsymbol{K}_{\Sigma\Sigma}^{-1}\boldsymbol{K}_{\Sigma\theta} \end{bmatrix}^{\text{hr}} \begin{bmatrix} \Delta\boldsymbol{q} \\ \Delta\boldsymbol{\Theta} \end{bmatrix} = \begin{bmatrix} \boldsymbol{R}_{\varphi} - \boldsymbol{K}_{\varphi\Sigma}{\boldsymbol{K}_{\Sigma\Sigma}}^{-1}\boldsymbol{R}_{\Sigma} \\ \boldsymbol{R}_{\theta} - \boldsymbol{K}_{\theta\Sigma}{\boldsymbol{K}_{\Sigma\Sigma}}^{-1}\boldsymbol{R}_{\Sigma} \end{bmatrix}^{\text{hr}} \tag{3.86}$$

to be solved with respect to $\Delta\boldsymbol{q}$ and $\Delta\boldsymbol{\Theta}$. In a last step, (3.85) can be used to obtain the stress update $\Delta\boldsymbol{\Sigma}$.

Remark 3: The proposed second order accurate time integration scheme can be replaced by a more simple implicit Euler approach, to which we refer to as endpoint rule. Therefore, we evaluate all corresponding terms at time t_{n+1} which will be used subsequently for quasi-static problems.

3.6 Numerical examples

In this section we present several quasi-static examples to demonstrate the applicability and performance of the Hellinger-Reissner type variational principle for fully coupled non-linear thermoelasticity. Additionally, we deal with a transient problem to investigate both, the Hellinger-Reissner as well as the displacement-temperature based approach, within the context of thermoelastodynamics.

The material behavior within the present examples is assumed to be governed by the proposed Mooney-Rivlin material law. The corresponding material parameters are specified separately in each example using SI-units. Note that the chosen values for the material parameters do not correspond to any physical material, i.e. the examples presented in this section are of purely academic nature.

3.6.1 Patch test

In order to demonstrate the applicability of the Hellinger-Reissner approach we start with a simple patch test example. To be specific, a block of size $0.5 \times 0.5 \times 0.5$ is loaded by a uniform pressure load of $p = 5 \times 10^6$ on the upper surface, whereas its lower surface is clamped, such that the body can expand in horizontal direction, see Figure 3.2. The block consists of $3 \times 3 \times 3$ quadratic B-spline elements.

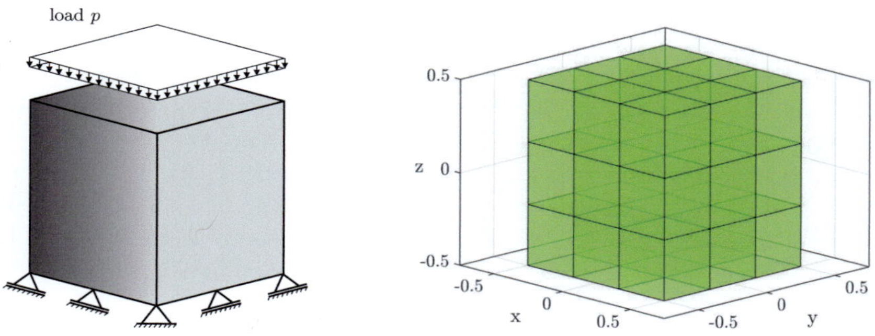

Figure 3.2: Patch test: Boundary conditions (left) and reference configuration (right).

The mechanical parameters for the Mooney-Rivlin material law are given by $\alpha = 15/13 \times 10^5$, $\beta = 10/13 \times 10^5$ and $\kappa = 25/3 \times 10^5$, which correspond to a Young's modulus of $E = 10^6$ and to a Poisson's ratio of $\nu = 0.3$. Moreover, the parameters for the thermal material behavior are specified by $c = 1830$, $\gamma = 0.22333$ and $K = K_{\text{ref}} = 0.55$. To obtain a quasi-static solution, the mass density is set to zero, i.e. only Γ^{h} within the thermal field depends on the differentiation with respect to time.

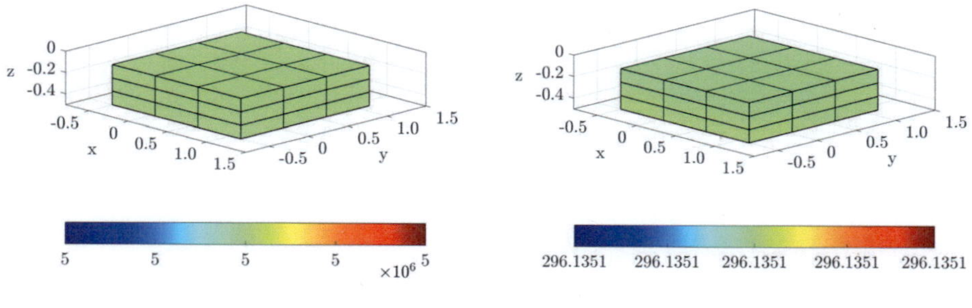

Figure 3.3: Patch test: Von Mises stress distribution (left) and temperature distribution (right) for the Hellinger-Reissner approach.

Figure 3.3 (left) shows the von Mises stress distribution of the solution for the applied uniform pressure load on the upper surface. The initial temperature is set to 293.15. Due to the applied linear expansion coefficient, the body heats up, as shown in Figure 3.3 (right). Nearly perfect and uniform results are obtained for the stress and temperature distributions. Therein, the color map for the von Mises stress distribution is in range of $5 \times 10^6 \pm 10^{-5}$, whereas a range wide of $\pm 2 \times 10^{-11}$ is used for the temperature distribution. Note that similar results are obtained for the displacement-temperature based approach.

3.6.2 Notched bar

Next, we consider a further quasi-static example to investigate the convergence of the Hellinger-Reissner approach in comparison to the displacement-temperature based approach. In particular, a rectangular bar of size $10 \times 4.8 \times 1$ is clamped at its ends, i.e. temporal Dirichlet boundary conditions are applied. Both sides are moved apart in longitudinal direction with a uniform incremental step size of $\Delta u_x = 0.1$, whereas the displacement in the other directions is fixed, see Figure 3.4. The bar is provided with a notch of approximately 1/3 of the width at the middle, see also Hesch & Betsch [38] and Holzapfel and Simo [47] for a similar setting.

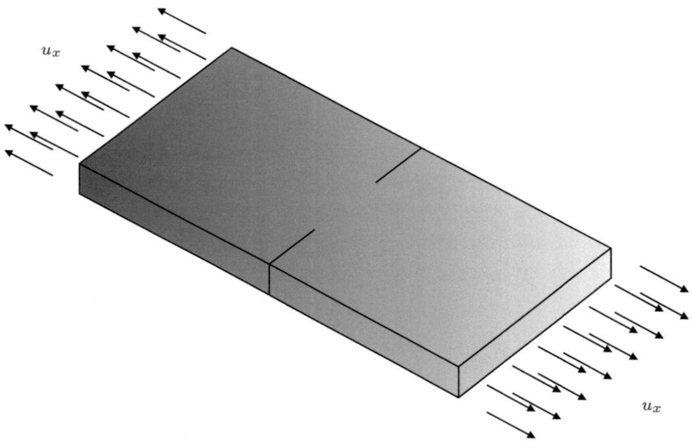

Figure 3.4: Notched bar: Boundary conditions.

The Mooney-Rivlin material parameters are given as follows, for the mechanical field the parameters take the values $\alpha = 1500/13$, $\beta = 1000/13$ and $\kappa = 2500/3$, whereas the setting for the thermal field reads $\theta_0 = 293.15$, $c = 1830$, $\gamma = 2.2333 \times 10^{-2}$ and $K = K_{\text{ref}} = 0.55$. The corresponding Young's modulus is given by $E = 1000$ and the Poisson's ratio by $\nu = 0.3$.

For the numerical simulations, a mesh consisting of $32 \times 12 \times 2$ quadratic B-spline elements with 5936 thermal and mechanical degrees of freedom is used as well as a two level locally refined mesh consisting of overall 5808 elements with in total 18982 degrees of freedom, see Figure 3.5.

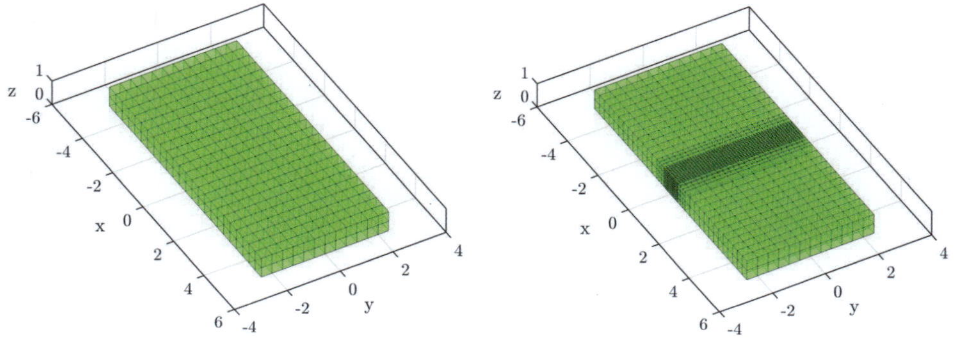

Figure 3.5: Notched bar: Reference configuration. Unrefined mesh (left) and two level refined mesh (right).

Figure 3.6 (left) shows the von Mises stress distribution of the deformed geometry at a displacement of $u_x = 3$. Additionally, the corresponding temperature distribution is plotted in Figure 3.6 (right). Due to large strains and associated stresses, respectively, we obtain the largest cooling down in the region around the notch.

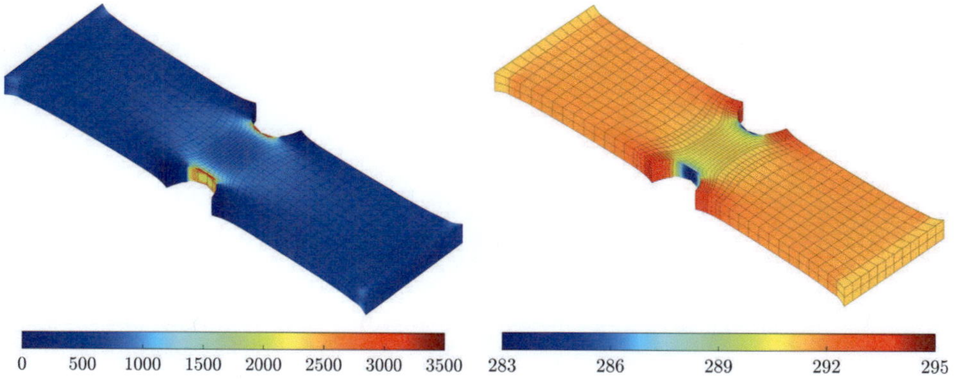

Figure 3.6: Notched bar: Von Mises stress distribution (left) and temperature distribution (right). Results for a displacement of $u = 3$ using the Hellinger-Reissner approach and the refined mesh are depicted.

Eventually, the load-deflection curve is plotted in Figure 3.7 for the Hellinger-Reissner as well as the displacement-temperature based approach and both meshes. For both approaches and both mesh resolutions almost identical results are obtained. Thus, we assume a nearly converged solution for the given meshes.

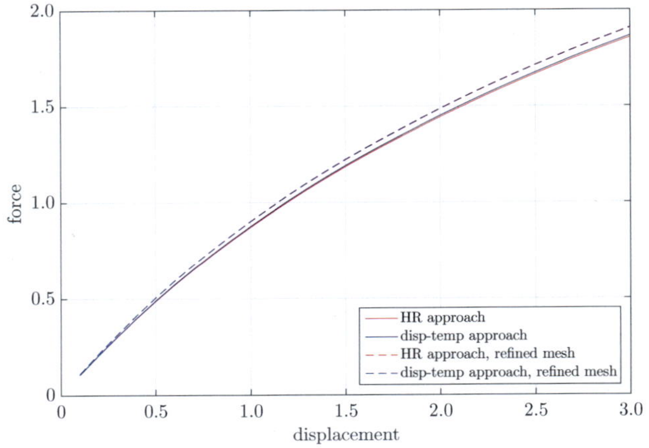

Figure 3.7: Notched bar: Load-deflection curves for the Hellinger-Reissner and the displace ment-temperature based approach using the unrefined as well as the refined mesh.

3.6.3 L-shaped block

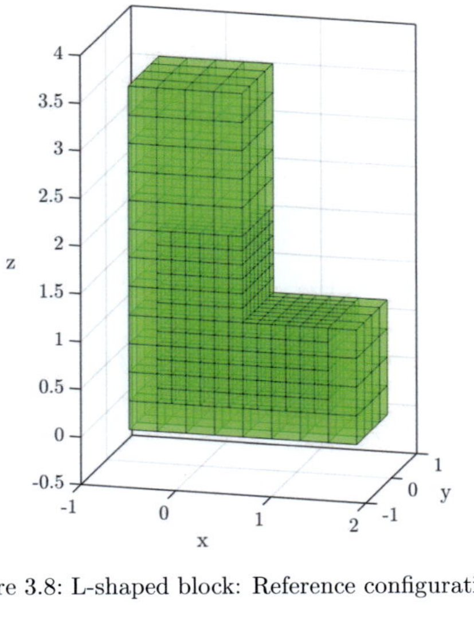

Figure 3.8: L-shaped block: Reference configuration.

Eventually, we deal with a transient thermoelastic problem consisting a L-shape block of size $2.4 \times 1.2 \times 3.6$ as depicted in Figure 3.8, see also Hesch & Betsch [38]. The L-shape is discretized with 256 quadratic B-spline elements on level zero. Additionally, a one level hierarchical refinement is applied to the region where we expect peak stresses due to the notch of the L-shape, see Figure 3.8. In total we obtain 1012 elements with overall 6792 thermal and mechanical degrees of freedom.

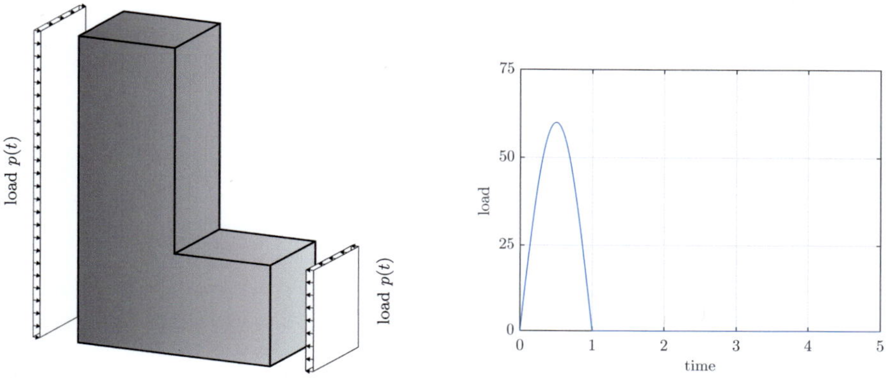

Figure 3.9: L-shaped block: Boundary conditions (left) and load over time (right).

The Mooney-Rivlin material parameters are given by $\alpha = 114.9425$, $\beta = 57.4713$ and $\kappa = 3333.3333$, which correspond to a Young's modulus of $E = 1000$ and to a Poisson's ratio of $\nu = 0.45$. The mass density is set to $\rho = 0.5$. Moreover, the thermal material setting reads $\theta_0 = 293.15$, $c = 1830$, $\gamma = 2.2333 \cdot 10^{-4}$ and $K_{\text{ref}} = 0.15$ with $\omega = 0.004$.

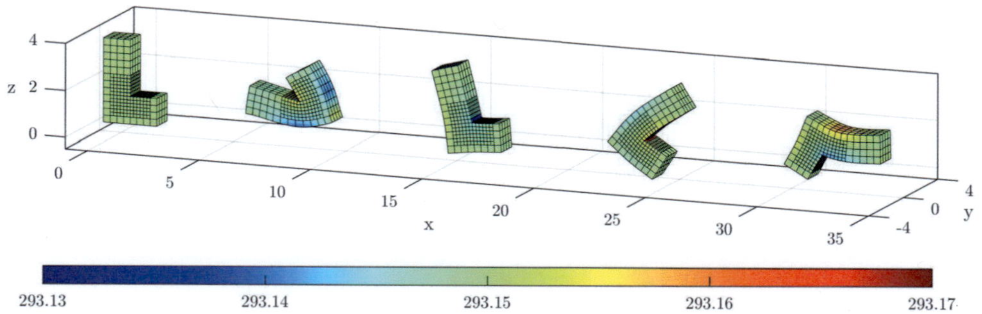

Figure 3.10: L-shaped block: Snapshots of the deformed configuration at different times $t = [0,\ 0.46,\ 0.71,\ 0.96,\ 1.21]$.

The body is subjected to a temporal traction load $p(t) = p_{\max} \sin(\pi t)$ during the time interval $t \in [0,\ 1]$ with a peak load of $p_{\max} = 60$, see Figure 3.9. Afterwards, the body moves freely in space for the time interval $t \in [1,\ 5]$.

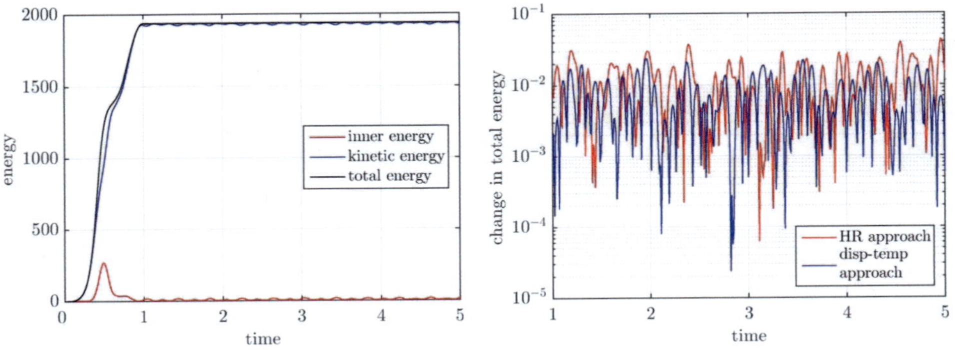

Figure 3.11: L-shaped block: Energy plotted over time (left) and change in total energy over the time interval $t \in [1,\ 5]$ (right).

The motion of the body and the associated temperature distribution is illustrated in Figure 3.10 using a time step size of $\Delta t = 0.01$. For both, the Hellinger-Reissner as well as the displacement-temperature based approach, the applied time step size is sufficiently small to conserve nearly the total energy using the proposed mid-point type time integration scheme, see Figure 3.11. As shown in Figure 3.10, the body turns while the load is fixed in its initial direction. This leads to the depicted energy curve during the load phase. In addition, the angular momentum is plotted over time in Figure 3.12. As expected, the components of angular momentum remain preserved for the considered problem and time integration scheme.

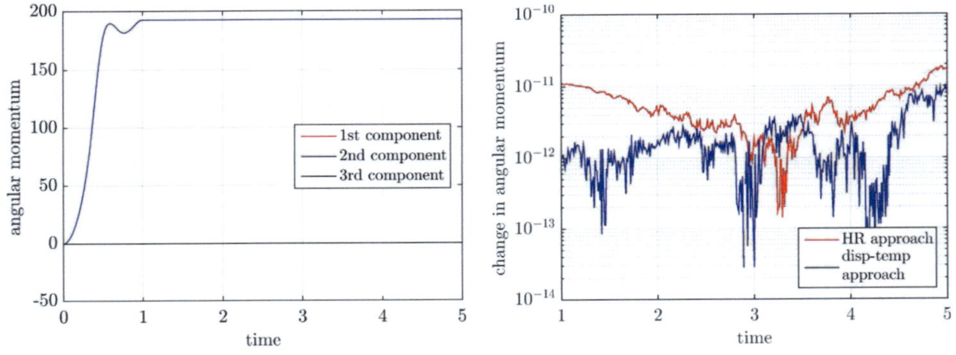

Figure 3.12: L-shaped block: Angular momentum plotted over time (left) and change in the second component of angular momentum over the time interval $t \in [1,\ 5]$ (right).

To sum up, the accuracy and performance of the Hellinger-Reissner approach are comparable to the displacement-temperature based approach. The implementation is slightly more complicated and requires a condensation procedure to avoid an unnecessarily high computational effort during the solving of the global system. However, the proposed Hellinger-Reissner type variational principle for thermomechanical systems represents a novel and alternative formulation and provides a direct access to the stresses, which leads to high flexibility for the modeling of multi-field problems.

4 Large strain phase-field approach to fracture mechanics

Variationally consistent formulated phase-field methods to fracture are able to predict complex three dimensional crack patterns, see, among many other, Miehe et al. [64, 66], Hesch & Weinberg [45] and Hesch et al. [42, 44]. In this chapter we deal with a fourth order phase-field approach to brittle fracture within a fully non-linear framework. In particular, the usage of the extended kinematic set as proposed in the previous chapter allows us to introduce a novel formulation based on an anisotropic split of the principle invariants in order to account for the different behavior of fracture in tension and compression. This formulation satisfies the polyconvexity criteria in the sense of Ball [8] for the mechanical field independent of the current state of fracture, cf. Hesch et al. [43]. Moreover, a modification of the degradation of strain energy is investigated which improves along with the higher order regularization of the crack topology, the accuracy and convergence rates of the numerical solution, see Borden [11], Borden et al. [12] and Weinberg & Hesch [98]. For the spatial discretization of the fourth order phase-field model, the B-spline based approximations are required to be at least C^1-continuous and the hierarchical refinement procedure enables the investigation of fracture problems in a three dimensional environment. Eventually, a series of numerical examples is extensively investigated to demonstrate the accuracy and performance of the proposed framework in comparison to previously established methods.

4.1 Configuration and kinematics

Similar to chapter 3, the body of interest is given in its reference configuration of the bounded domain $\mathcal{B} \subset \mathbb{R}^d$ with $d = 3$. Again, a sufficiently smooth non-linear deformation mapping

$$\boldsymbol{\varphi}(\boldsymbol{X}, t) : \mathcal{B}_0 \times \mathcal{T} \rightarrow \mathbb{R}^d \tag{4.1}$$

is introduced to map a material point $\boldsymbol{X}$ in its reference configuration to its current position $\boldsymbol{x} = \boldsymbol{\varphi}(\boldsymbol{X}, t)$ at time $t \in \mathcal{I} = [0, T]$. In addition to the mechanical field, we assume the existence of a sufficiently smooth phase-field to characterize diffusive crack modeling inside the elastic body. The phase-field is described by an order parameter

$$s(\boldsymbol{X}, t) : \mathcal{B}_0 \times \mathcal{T} \rightarrow [0, 1], \tag{4.2}$$

where the value $s = 0$ refers to the undamaged and $s = 1$ to the fully broken state of the material. Now, the unknowns $\{\boldsymbol{\varphi}, s\}$ from a configuration space in $\mathbb{R}^{d+1}$, representing the primal degrees of freedom to be found for all times of interest.

For the reminder, we introduce an extended kinematic set given by the deformation gradient

$$\boldsymbol{F} = \nabla(\boldsymbol{\varphi}), \tag{4.3}$$

its co-factor

$$\boldsymbol{H} = \frac{1}{2}(\boldsymbol{F} \times \boldsymbol{F}) \tag{4.4}$$

and its determinant

$$J = \frac{1}{6}(\boldsymbol{F} \times \boldsymbol{F}) : \boldsymbol{F}. \tag{4.5}$$

4.2 Energy balance and constitutive theory for brittle fracture

We state the global energy balance equation for the system under consideration as

$$\dot{T}(t) + \dot{W}^{\mathrm{e}}(t) + \dot{W}^{\mathrm{cr}}(t) = P^{\mathrm{ext}}(t), \tag{4.6}$$

where thermal effects as discussed in chapter 3 are omitted. In addition to the kinetic energy T and the external mechanical power P^{ext}, W^{e} and W^{cr} denote the stored elastic strain energy and the fracture energy, respectively. Following the classical brittle fracture approach of Griffith [34] and Irwin [51], the crack initiates or continues upon the attainment of a critical local fracture energy density g_{c}. Thus, the fracture energy within an arising internal sharp crack interface $\partial\mathcal{B}_0^{\mathrm{cr}}(t) \subset \mathbb{R}^{d-1}$ reads

$$W^{\mathrm{cr}}(t) = \int\limits_{\partial\mathcal{B}_0^{\mathrm{cr}}(t)} g_{\mathrm{c}} \mathrm{d}A. \tag{4.7}$$

Now, the mechanical energy balance equation in material description can be written as

$$\frac{\mathrm{d}}{\mathrm{d}t}\int\limits_{\mathcal{B}_0}\left(\frac{1}{2}\rho_0\dot{\boldsymbol{\varphi}}\cdot\dot{\boldsymbol{\varphi}} + w^{\mathrm{e}}\right)\mathrm{d}V + \frac{\mathrm{d}}{\mathrm{d}t}\int\limits_{\partial\mathcal{B}_0^{\mathrm{cr}}(t)} g_{\mathrm{c}}\mathrm{d}A = \int\limits_{\partial\mathcal{B}_0}\boldsymbol{T}\cdot\dot{\boldsymbol{\varphi}}\mathrm{d}A + \int\limits_{\mathcal{B}_0}\boldsymbol{B}\cdot\dot{\boldsymbol{\varphi}}\mathrm{d}V, \tag{4.8}$$

where w^{e} is the elastic strain energy density defined with respect to the reference volume. Following the explanations in section 3.2, we obtain

$$\int\limits_{\mathcal{B}_0}\dot{w}^{\mathrm{e}}\mathrm{d}V + \frac{\mathrm{d}}{\mathrm{d}t}\int\limits_{\partial\mathcal{B}_0^{\mathrm{cr}}(t)} g_{\mathrm{c}}\mathrm{d}A = \int\limits_{\mathcal{B}_0}(\boldsymbol{P}^{\mathrm{e}} : \dot{\boldsymbol{F}} + (\mathrm{Div}(\boldsymbol{P}^{\mathrm{e}}) + \boldsymbol{B} - \rho_0\ddot{\boldsymbol{\varphi}})\cdot\dot{\boldsymbol{\varphi}})\mathrm{d}V \tag{4.9}$$

and satisfying the balance of linear momentum yields

$$\int\limits_{\mathcal{B}_0}\dot{w}^{\mathrm{e}}\mathrm{d}V + \frac{\mathrm{d}}{\mathrm{d}t}\int\limits_{\partial\mathcal{B}_0^{\mathrm{cr}}(t)} g_{\mathrm{c}}\mathrm{d}A = \int\limits_{\mathcal{B}_0}\boldsymbol{P}^{\mathrm{e}} : \dot{\boldsymbol{F}}\mathrm{d}V. \tag{4.10}$$

Thus, the rate of change of the elastic strain energy plus the rate of change of the fracture energy equals the elastic stress power.

The second law of thermodynamics as postulated by (3.16) simplifies for isothermal and adiabatic processes (as considered in this chapter) to

$$D^{\text{int}} = \theta \dot{\eta} \geq 0, \tag{4.11}$$

such that a change in entropy is purely caused by internal dissipation. For the considered system, (3.20) becomes

$$D^{\text{int}} = \boldsymbol{P}^{\text{e}} : \dot{\boldsymbol{F}} - \dot{w}^{\text{e}} \geq 0 \tag{4.12}$$

and insertion into (4.10) yields

$$\frac{\text{d}}{\text{d}t} \int\limits_{\partial \mathcal{B}_0^{\text{cr}}(t)} g_{\text{c}} \text{d}A \geq 0, \tag{4.13}$$

i.e. the second law of thermodynamics prohibits to transfer fracture energy back into the mechanical field.

4.2.1 Phase field model

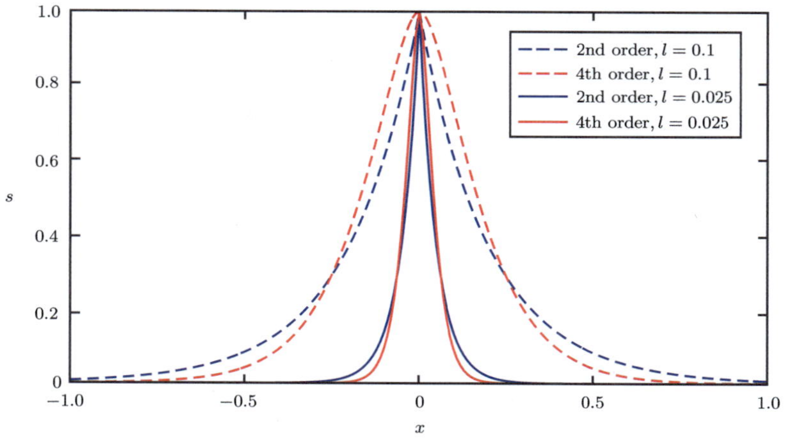

Figure 4.1: Analytical solution of the second order phase-field $s(x) = e^{\frac{-|x|}{2l}}$ and fourth order phase-field $s(x) = e^{\frac{-|x|}{l}} \left(1 + \frac{|x|}{l}\right)$ for different values of the length scale parameter.

Since the numerical evaluation of sharp crack interfaces is not suitable within a standard finite element framework, a regularized crack interface using a specific regularization profile $\gamma(s)$ is introduced, such that the fracture energy is approximated by

$$\int_{\partial \mathcal{B}^{cr}(t)} g_c \mathrm{d}A \approx \int_{\mathcal{B}_0} g_c \gamma(s) \mathrm{d}V. \tag{4.14}$$

For the regularization zone, a higher order regularization of the crack topology is utilized to obtain better accuracy and convergence rates of the numerical solution, see Borden et al. [12] for a detailed discussion. Hence, we apply a fourth order approach for the crack density functional

$$\bar{\gamma}(s, \nabla(s), \Delta(s)) = \frac{1}{4l}s^2 + \frac{l}{2}\nabla(s) \cdot \nabla(s) + \frac{l^3}{4}\Delta(s)\Delta(s), \tag{4.15}$$

where the intrinsic length scale parameter l controls the regularization zone. Note that this parameter has substantial impact on the crack initialization and can be regarded as material parameter, see Miehe et al. [65]. In Figure 4.1 the analytical solution of the corresponding Euler equation of the fourth order approach in the one dimensional case is compared to the solution of the Euler equation of the second order crack density functional

$$\hat{\gamma}(s, \nabla(s)) = \frac{1}{2l}s^2 + \frac{l}{2}\nabla(s) \cdot \nabla(s). \tag{4.16}$$

Note that there is no natural choice of differential equation to model the crack density functional, for a survey see Weinberg & Hesch [98] and the references therein. Although the fourth order approach is superior against the second order approach in terms of accuracy, it additionally minimizes the global size of crack interfaces (see Cahn & Hilliard [15]) and is therefore not necessarily suitable to handle effects like transient fragmentation and crack branching.

4.2.2 Elastic strain energy

A fully isotropic formulation for the degradation of a strain energy density function due to fracture requires a monotonically decreasing function $g(s)$ with properties $g(0) = 1$, $g(1) = 0$ and $g'(1) = 0$. The most simplest constitutive assumption reads $\hat{g}(s) = (1-s)^2$ and we obtain

$$w^{\mathrm{e}}(\boldsymbol{F}, s) = \hat{g}(s) w(\boldsymbol{F}). \tag{4.17}$$

A physically more reasonable formulation relies on a decomposition into tensile and compressive components, assuming that fracture requires a local state of tension. Therefore, we propose an additive anisotropic split related to the principal invariants of the right Cauchy-Green tensor, which also necessitates a deviatoric split. Otherwise the volumetric part within the first and second invariant might contradict the anisotropic split of the third invariant, resulting in an unphysical behavior of the formulation. The tensile and the compressive components of the isochoric contribution of the first invariant, i.e. $I_{\bar{C}}^{+}$ and $I_{\bar{C}}^{-}$, are given as

$$I_{\bar{C}}^{\pm}(I_C(\boldsymbol{F}), J) = J^{-2/3}\left(\frac{(I_C(\boldsymbol{F}) - d) \pm |I_C(\boldsymbol{F}) - d|}{2} + d\right) \tag{4.18}$$

and for the corresponding components of the second invariant follow analogously

$$II_{\bar{C}}^{\pm}(II_C(\boldsymbol{H}), J) = J^{-4/3}\left(\frac{(II_C(\boldsymbol{H}) - d) \pm |II_C(\boldsymbol{H}) - d|}{2} + d\right), \tag{4.19}$$

where $I_C(\boldsymbol{F}) = \boldsymbol{F} : \boldsymbol{F}$ and $II_C(\boldsymbol{H}) = \boldsymbol{H} : \boldsymbol{H}$, see also section 3.3.1. In addition, for the Jacobian determinant we obtain

$$J^{\pm}(J) = \frac{(J-1) \pm |J-1|}{2} + 1. \tag{4.20}$$

Now, the elastic strain energy density function of a standard Mooney-Rivlin material law can be defined as

$$\bar{w}^{\mathrm{e}}(\boldsymbol{F}, \boldsymbol{H}, J, s) = \bar{w}^{\mathrm{e,iso}}(I_{\bar{C}}^{\pm}(I_C(\boldsymbol{F}), J), II_{\bar{C}}^{\pm}(II_C(\boldsymbol{H}), J), s) + \bar{w}^{\mathrm{e,vol}}(J^{\pm}(J), s), \tag{4.21}$$

where the isochoric contribution reads

$$\bar{w}^{\mathrm{e,iso}}(I_{\bar{C}}^{\pm}, II_{\bar{C}}^{\pm}, s) = \alpha\bar{g}(s)(I_{\bar{C}}^{+} - d) + \alpha(I_{\bar{C}}^{-} - d) + \beta\bar{g}(s)(II_{\bar{C}}^{+\,3/2} - d^{3/2}) + \beta(II_{\bar{C}}^{-\,3/2} - d^{3/2}) \tag{4.22}$$

and the volumetric contribution is given as

$$\bar{w}^{\mathrm{e,vol}}(J^{\pm}, s) = \frac{\kappa}{2}(\bar{g}(s)(J^{+} - 1))^2 + \frac{\kappa}{2}(J^{-} - 1)^2. \tag{4.23}$$

Therein, $\hat{g}(s) = (1-s)^2$ is replaced by a cubic degradation function

$$\bar{g}(s) = a_{\mathrm{g}}((1-s)^3 - (1-s)^2) - 2(1-s)^3 + 3(1-s)^2 \tag{4.24}$$

with $a_{\mathrm{g}} \geq 0$ and properties $\bar{g}(0) = 1$, $\bar{g}(1) = 0$, $\bar{g}'(0) = -a_{\mathrm{g}}$ and $\bar{g}'(1) = 0$. Note that we obtain again the quadratic degradation function by setting $a_{\mathrm{g}} = 2$. However, in order to avoid that relatively small strains even far from a crack noticeably contribute to the computed fracture energy, we apply the proposed cubic degradation function along with small values of a_{g}, cf. Borden [11]. In Figure 4.2, the cubic degradation function is compared for different values of the modeling parameter a_{g}. In order to avoid a full degradation of strain energy at the fully broken state, i.e. $s = 1$, we introduce a small positive parameter $k_{\mathrm{g}} \approx 0$ such that $\bar{g}(1) = k_{\mathrm{g}}$, see also Miehe et al. [66]. As shown in Hesch et al. [43], this ensures that (4.21) is polyconvex with respect to $\boldsymbol{F}$, $\boldsymbol{H}$ and J taking into account arbitrary values of the phase-field parameter.

An alternative formulation of the volumetric contribution to the strain energy density function using a multiplicative anisotropic split of the Jacobian determinant reads

$$\hat{w}^{\mathrm{e,vol}}(J^{\pm}, s) = \frac{\kappa}{2}((J^{+})^{\bar{g}(s)}J^{-} - 1)^2. \tag{4.25}$$

However, by applying this formulation polyconvexity with respect to $\boldsymbol{F}$, $\boldsymbol{H}$ and J depends on the chosen set of positive material parameters.

Remark 4: The proposed elastic strain energy function (4.21) describing a Mooney-Rivlin material can be reduced to a Neo-Hookean material law by neglecting the latter both terms in (4.22). Typically, a standard Neo-Hookean material model is formulated using the Lamé parameters λ and μ, e.g. as

$$w(\boldsymbol{F}, J) = \frac{\mu}{2}(I_C - d) + \frac{\lambda}{2}(J-1)^2 - \mu \ln(J), \tag{4.26}$$

whereas a corresponding alternative constitutive model using an isochoric split reads

$$\bar{w}(\boldsymbol{F}, J) = \frac{\mu}{2}(I_{\bar{C}} - d) + \frac{\kappa}{2}(J-1)^2. \tag{4.27}$$

Then, the relation $\kappa = (3\lambda + 2\mu)/3$ (3D case) and $\kappa = \lambda + \mu$ (2D case) is used for the bulk modulus, where only for the three dimensional case an exact correlation with the Lamé parameters is given. Moreover, $\mu = 2(\alpha + \beta)$ is used to achieve nearly comparability to the Mooney-Rivlin material law.

Remark 5: The above formulations are not valid for two dimensional problems, but can be adapted without changing the proposed framework. Two dimensional problems enables us to investigate the phase-field approach using high mesh resolutions without possible influence of local refinements.

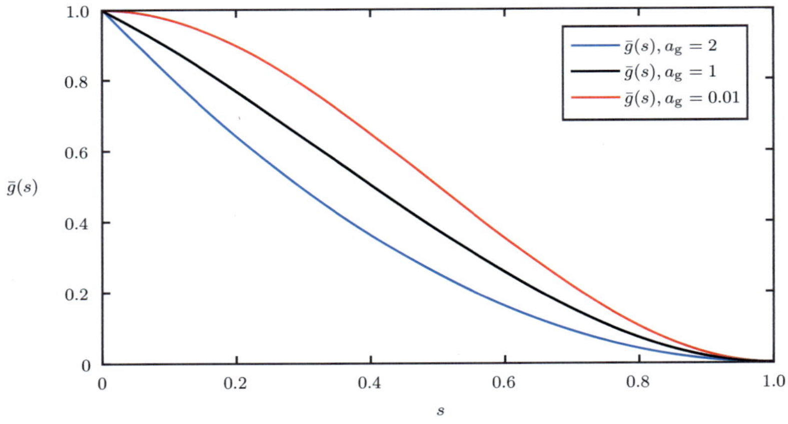

Figure 4.2: Cubic degradation function $\bar{g}(s)$ plotted over the phase-field parameter s for $a_g = 2$, $a_g = 1$ and $a_g = 0.01$.

4.2.3 Conjugate stresses, phase-field driving force and Hessian operator

Similar to the thermomechanical formulation in section 3.3.2, we introduce work conjugate stresses to the extended kinematic set $\{\boldsymbol{F}, \boldsymbol{H}, J\}$ as follows

$$\boldsymbol{\Sigma}^{\mathrm{e}}_F(\boldsymbol{F}, J, s) = \frac{\partial \bar{w}^{\mathrm{e}}}{\partial \boldsymbol{F}}, \qquad \boldsymbol{\Sigma}^{\mathrm{e}}_H(\boldsymbol{H}, J, s) = \frac{\partial \bar{w}^{\mathrm{e}}}{\partial \boldsymbol{H}} \qquad \text{and} \qquad \Sigma^{\mathrm{e}}_J(\boldsymbol{F}, \boldsymbol{H}, J, s) = \frac{\partial \bar{w}^{\mathrm{e}}}{\partial J}, \tag{4.28}$$

whereas the driving force to the phase-field equation is obtained via

$$H(\boldsymbol{F}, \boldsymbol{H}, J, s) = \frac{\partial \bar{w}^{\mathrm{e}}}{\partial s}. \tag{4.29}$$

Hence, the first variation of the strain energy density function given by (4.21) with respect to the primal variables reads

$$\begin{aligned} &\mathrm{D}\bar{w}^{\mathrm{e}}[\mathrm{D}\boldsymbol{F}[\delta\boldsymbol{\varphi}], \mathrm{D}\boldsymbol{H}[\delta\boldsymbol{\varphi}], \mathrm{D}J[\delta\boldsymbol{\varphi}], \delta s] \\ &\quad = \boldsymbol{\Sigma}^{\mathrm{e}}_{\boldsymbol{F}} : \mathrm{D}\boldsymbol{F}[\delta\boldsymbol{\varphi}] + \boldsymbol{\Sigma}^{\mathrm{e}}_{\boldsymbol{H}} : \mathrm{D}\boldsymbol{H}[\delta\boldsymbol{\varphi}] + \Sigma^{\mathrm{e}}_J\, \mathrm{D}J[\delta\boldsymbol{\varphi}] + H\delta s \\ &\quad = (\boldsymbol{\Sigma}^{\mathrm{e}}_F + \boldsymbol{\Sigma}^{\mathrm{e}}_H \times \boldsymbol{F} + \Sigma^{\mathrm{e}}_J \boldsymbol{H}) : \nabla(\delta\boldsymbol{\varphi}) + H\delta s\,. \end{aligned} \tag{4.30}$$

As before, the relationship between the conjugate stresses and the first Piola-Kirchhoff stress tensor is given as follows

$$\boldsymbol{P}^{\mathrm{e}} = \boldsymbol{\Sigma}^{\mathrm{e}}_F + \boldsymbol{\Sigma}^{\mathrm{e}}_H \times \boldsymbol{F} + \Sigma^{\mathrm{e}}_J \boldsymbol{H}. \tag{4.31}$$

Eventually, the linearization with respect to the primal variables reads

$$\begin{aligned} \mathrm{D}^2\bar{w}^{\mathrm{e}}[\delta\boldsymbol{\varphi}, \delta s, \Delta\boldsymbol{\varphi}, \Delta s] &= \begin{bmatrix} \nabla(\delta\boldsymbol{\varphi}) : & \delta s \end{bmatrix} \begin{bmatrix} \mathrm{D}\boldsymbol{P}^{\mathrm{e}}[\nabla(\Delta\boldsymbol{\varphi}), \Delta s] \\ \mathrm{D}H[\nabla(\Delta\boldsymbol{\varphi}), \Delta s] \end{bmatrix} \\ &= \begin{bmatrix} \nabla(\delta\boldsymbol{\varphi}) : \\ \nabla(\delta\boldsymbol{\varphi}) \times \boldsymbol{F} : \\ \nabla(\delta\boldsymbol{\varphi}) : \boldsymbol{H} \\ \delta s \end{bmatrix}^{\mathrm{T}} [\mathbb{H}_{\bar{w}^{\mathrm{e}}}] \begin{bmatrix} : \nabla(\Delta\boldsymbol{\varphi}) \\ : \nabla(\Delta\boldsymbol{\varphi}) \times \boldsymbol{F} \\ \nabla(\Delta\boldsymbol{\varphi}) : \boldsymbol{H} \\ \Delta s \end{bmatrix} \\ &\quad + (\boldsymbol{\Sigma}^{\mathrm{e}}_{\boldsymbol{H}} + \Sigma^{\mathrm{e}}_J \boldsymbol{F}) : (\nabla(\delta\boldsymbol{\varphi}) \times \nabla(\Delta\boldsymbol{\varphi})), \end{aligned} \tag{4.32}$$

where the Hessian operator is given as

$$[\mathbb{H}_{\bar{w}^{\mathrm{e}}}] = \begin{bmatrix} \frac{\partial^2 \bar{w}^{\mathrm{e}}}{\partial \boldsymbol{F} \partial \boldsymbol{F}} & \frac{\partial^2 \bar{w}^{\mathrm{e}}}{\partial \boldsymbol{F} \partial \boldsymbol{H}} & \frac{\partial^2 \bar{w}^{\mathrm{e}}}{\partial \boldsymbol{F} \partial J} & \frac{\partial^2 \bar{w}^{\mathrm{e}}}{\partial \boldsymbol{F} \partial s} \\ \frac{\partial^2 \bar{w}^{\mathrm{e}}}{\partial \boldsymbol{H} \partial \boldsymbol{F}} & \frac{\partial^2 \bar{w}^{\mathrm{e}}}{\partial \boldsymbol{H} \partial \boldsymbol{H}} & \frac{\partial^2 \bar{w}^{\mathrm{e}}}{\partial \boldsymbol{H} \partial J} & \frac{\partial^2 \bar{w}^{\mathrm{e}}}{\partial \boldsymbol{H} \partial s} \\ \frac{\partial^2 \bar{w}^{\mathrm{e}}}{\partial J \partial \boldsymbol{F}} & \frac{\partial^2 \bar{w}^{\mathrm{e}}}{\partial J \partial \boldsymbol{H}} & \frac{\partial^2 \bar{w}^{\mathrm{e}}}{\partial J \partial J} & \frac{\partial^2 \bar{w}^{\mathrm{e}}}{\partial J \partial s} \\ \frac{\partial^2 \bar{w}^{\mathrm{e}}}{\partial s \partial \boldsymbol{F}} & \frac{\partial^2 \bar{w}^{\mathrm{e}}}{\partial s \partial \boldsymbol{H}} & \frac{\partial^2 \bar{w}^{\mathrm{e}}}{\partial s \partial J} & \frac{\partial^2 \bar{w}^{\mathrm{e}}}{\partial s \partial s} \end{bmatrix}. \tag{4.33}$$

The above first and second derivatives with respect to the three strain measures are tedious but straightforward (cf. section 3.3.2), whereas the derivatives with respect to the phase-field simply rely on the derivative of the degradation function.

4.3 Variational formulation

Utilizing the above derivations we can recast (4.9) as

$$\int\limits_{\mathcal{B}_0} \left((\dot{\boldsymbol{\pi}} - \mathrm{Div}(\boldsymbol{P}^{\mathrm{e}}) - \boldsymbol{B}) \cdot \dot{\boldsymbol{\varphi}} + \left(H + g_{\mathrm{c}} \frac{\partial \bar{\gamma}}{\partial s} \right) \dot{s} + \frac{\partial \bar{\gamma}}{\partial \nabla(s)} \cdot \nabla(\dot{s}) + \frac{\partial \bar{\gamma}}{\partial \Delta(s)} \Delta(\dot{s}) \right) \mathrm{d}V = 0, \tag{4.34}$$

where $\int_{\mathcal{B}_0} g_c \dot{\bar{\gamma}} \mathrm{d}V$ can also be regarded as global dissipation term, cf. (4.13). The variation of (4.34) with respect to the rates of change of the primal variables along with integration by parts and divergence theorem leads to

$$\begin{aligned} G_{\varphi^e} &= \int_{\mathcal{B}_0} (\delta\boldsymbol{\varphi} \cdot \dot{\boldsymbol{\pi}} + \boldsymbol{P}^e : \nabla(\delta\boldsymbol{\varphi}))\mathrm{d}V - \int_{\mathcal{B}_0} \delta\boldsymbol{\varphi} \cdot \boldsymbol{B}\mathrm{d}V - \int_{\partial\mathcal{B}_0^T} \delta\boldsymbol{\varphi} \cdot \bar{\boldsymbol{T}}\mathrm{d}A = 0, \\ G_s &= \int_{\mathcal{B}_0} \left(\delta s \left(H + \frac{g_c}{2l} s \right) + g_c l \nabla(\delta s) \cdot \nabla(s) + \frac{g_c l^3}{2} \Delta(\delta s)\Delta(s) \right) \mathrm{d}V = 0. \end{aligned} \tag{4.35}$$

These equations have to hold for all $\delta\boldsymbol{\varphi} \in \mathcal{V}^{\varphi}$ and $\delta s \in \mathcal{V}^s$. Additionally, we enforce the constitutive relations (4.28) and (4.29) as well as the kinematic relation $\boldsymbol{\pi} = \rho_0 \dot{\boldsymbol{\varphi}}$ locally. The space of the virtual variation of the geometry is given by

$$\mathcal{V}^{\varphi} = \{\delta\boldsymbol{\varphi} \in \mathcal{H}^1(\mathcal{B}_0) | \delta\boldsymbol{\varphi} = \boldsymbol{0} \,\mathrm{on}\, \partial\mathcal{B}_0^{\varphi} \times \mathcal{T}\}, \tag{4.36}$$

whereas the corresponding space of the virtual variation of the phase-field is defined as

$$\mathcal{V}^{s} = \{\delta s \in \mathcal{H}^2(\mathcal{B}_0) | \delta s = 0 \,\mathrm{on}\, \partial\mathcal{B}_0^{s} \times \mathcal{T}\}, \tag{4.37}$$

i.e. the fully broken state on $\partial\mathcal{B}_0^s$ is treated as Dirichlet type constraint. Note that Miehe et al. [64], also demands $\dot{s} \geq 0$, which is equivalent to $\dot{\bar{\gamma}} \geq 0$ since $H \leq 0$, for thermodynamical consistency, avoiding a transfer of energy from the phase-field into the mechanical field. This prevents healing effects, which may be taken into account as well. Here, we slightly relax the condition given by (4.11) and restrict only the fully broken state, i.e. we allow for healing until the phase-field parameter reaches one, see Hesch & Weinberg [45].

4.4 Discrete setting

The finite element approximations of the deformed geometry $\boldsymbol{\varphi}$ and its variation $\delta\boldsymbol{\varphi}$ are defined as

$$\boldsymbol{\varphi}^{\mathrm{h}} = \sum_{A \in \mathcal{I}} R^A \boldsymbol{q}_A \qquad \text{and} \qquad \delta\boldsymbol{\varphi}^{\mathrm{h}} = \sum_{A \in \mathcal{I}} R^A \delta\boldsymbol{q}_A, \tag{4.38}$$

whereas we introduce corresponding approximations of the phase-field s and its variation δs as

$$s^{\mathrm{h}} = \sum_{A \in \mathcal{I}} R^A s_A \qquad \text{and} \qquad \delta s^{\mathrm{h}} = \sum_{A \in \mathcal{I}} R^A \delta s_A, \tag{4.39}$$

where s_A are the control variables for the phase-field. As before, we use the same B-spline based approximations for both fields, satisfying the required continuity $s^{\mathrm{h}} \in \mathcal{H}^2(\mathcal{B}_0)$ of the fourth order phase-field approach.

Beginning with the mechanical balance equation in weak form $(4.35)_1$, the semi-discrete counterpart is obtained in analogy to section 3.5.1. The first Piola-Kirchhoff stress tensor in semi-discrete setting reads

$$\boldsymbol{P}^{\mathrm{e,h}} = \boldsymbol{\Sigma}_F^{\mathrm{e,h}} + \boldsymbol{\Sigma}_H^{\mathrm{e,h}} \times \boldsymbol{F}^{\mathrm{h}} + \Sigma_J^{\mathrm{e,h}} \boldsymbol{H}^{\mathrm{h}}, \tag{4.40}$$

where the set of conjugate stresses is obtained via the semi-discrete version of the elastic strain energy density function $\bar{w}^{\mathrm{e,h}} = \bar{w}^{\mathrm{e,h}}(\boldsymbol{F}^{\mathrm{h}}, \boldsymbol{H}^{\mathrm{h}}, J^{\mathrm{h}}, s^{\mathrm{h}})$ as

$$\boldsymbol{\Sigma}_F^{\mathrm{e,h}} = \frac{\partial \bar{w}^{\mathrm{e,h}}}{\partial \boldsymbol{F}^{\mathrm{h}}}, \qquad \boldsymbol{\Sigma}_H^{\mathrm{e,h}} = \frac{\partial \bar{w}^{\mathrm{e,h}}}{\partial \boldsymbol{H}^{\mathrm{h}}} \qquad \text{and} \qquad \Sigma_J^{\mathrm{e,h}} = \frac{\partial \bar{w}^{\mathrm{e,h}}}{\partial J^{\mathrm{h}}}. \tag{4.41}$$

Thus, the semi-discrete counterpart of $(4.35)_1$ can be written as

$$\begin{aligned} G^{\mathrm{h}}_{\varphi^{\mathrm{e}}} &= \delta \boldsymbol{q}_A \cdot \left[M^{AB} \dot{\boldsymbol{v}}_B + \int_{\mathcal{B}_0} \boldsymbol{P}^{\mathrm{e,h}} \nabla(R^A) \mathrm{d}V - \boldsymbol{F}^{\mathrm{ext},A} \right] \\ &= \delta \boldsymbol{q}_A \cdot \boldsymbol{R}^A_{\varphi^{\mathrm{e}}}, \end{aligned} \tag{4.42}$$

cf. (3.65) within the thermomechanical formulation.

Next, we insert (4.39) into the weak form $(4.35)_2$. Then the corresponding semi-discrete counterpart reads

$$\begin{aligned} G^{\mathrm{h}}_s &= \delta s_A \left[\int_{\mathcal{B}_0} H^{\mathrm{h}} R^A \mathrm{d}V + \frac{g_c}{2l} K^{AB} s_B + g_c l K^{AB}_{\nabla} s_B + \frac{g_c l^3}{2} K^{AB}_{\Delta} s_B \right] \\ &= \delta s_A R^A_s, \end{aligned} \tag{4.43}$$

where the semi-discrete version of the phase-field driving force is obtained via

$$H^{\mathrm{h}} = \frac{\partial \bar{w}^{\mathrm{e,h}}}{\partial s^{\mathrm{h}}} \tag{4.44}$$

and the coefficients of the matrices $\boldsymbol{K}$, $\boldsymbol{K}_{\nabla}$ and $\boldsymbol{K}_{\Delta}$ are defined as

$$K^{AB} = \int_{\mathcal{B}_0} R^A R^B \mathrm{d}V, \quad K^{AB}_{\nabla} = \int_{\mathcal{B}_0} \nabla(R^A) \cdot \nabla(R^B) \mathrm{d}V \quad \text{and} \quad K^{AB}_{\Delta} = \int_{\mathcal{B}_0} \Delta(R^A) \Delta(R^B) \mathrm{d}V. \tag{4.45}$$

Eventually, we perform the temporal discretization for the considered fully coupled problem. Again, we subdivide the considered time interval $\mathcal{T}$ into a sequence of times, i.e. $t_0, \ldots, t_n, t_{n+1}, \ldots, T$ and assume the state variables at t_n, now given by $\boldsymbol{\varphi}^h_n$ and s^h_n are known. Then, the full-discrete setting of $(4.35)_1$ using the proposed mid-point evaluation reads

$$\begin{aligned} G^{\mathrm{h}}_{\varphi^{\mathrm{e}},n,n+1} &= \delta \boldsymbol{q}_A \cdot \left[M^{AB} \frac{\boldsymbol{v}_{B,n+1} - \boldsymbol{v}_{B,n}}{\Delta t} + \int_{\mathcal{B}_0} \nabla(R^A) \cdot \boldsymbol{P}^{\mathrm{e,h}}_{n,n+1} \mathrm{d}V - \boldsymbol{F}^{\mathrm{ext},A}_{n+1/2} \right] \\ &= \delta \boldsymbol{q}_A \cdot \boldsymbol{R}^A_{\varphi^{\mathrm{e}},n,n+1} \end{aligned} \tag{4.46}$$

and the full-discrete setting of $(4.35)_2$ takes the form

$$\begin{aligned} G^{\mathrm{h}}_{s,n,n+1} &= \delta s_A \left[\int_{\mathcal{B}_0} H^{\mathrm{h}}_{n,n+1} R^A \mathrm{d}V + \left[\frac{g_c}{2l} K^{AB} + g_c l K^{AB}_{\nabla} + \frac{g_c l^3}{2} K^{AB}_{\Delta} \right] s_{B,n+1/2} \right] \\ &= \delta s_A R^A_{s,n,n+1} \,. \end{aligned} \tag{4.47}$$

The above set of non-linear equations is supplemented by the mid-point type approximation of the kinematic relationship given by (3.78). Furthermore, the full-discrete phase-field driving force $H^{\mathrm{h}}_{n,n+1}$ is evaluated via the mid-point configuration of the primal variables given as

$$\boldsymbol{\varphi}^{\mathrm{h}}_{n+1/2} = \frac{\boldsymbol{\varphi}^{\mathrm{h}}_{n+1} + \boldsymbol{\varphi}^{\mathrm{h}}_{n}}{2} \quad \text{and} \quad s^{\mathrm{h}}_{n+1/2} = \frac{s^{\mathrm{h}}_{n+1} + s^{\mathrm{h}}_{n}}{2}. \tag{4.48}$$

Linearization

The global system to be solved within the Newton-Raphson iteration reads

$$\begin{bmatrix} \boldsymbol{K}_{\varphi^{\mathrm{e}}\varphi^{\mathrm{e}}} & \boldsymbol{K}_{\varphi^{\mathrm{e}}s} \\ \boldsymbol{K}_{s\varphi^{\mathrm{e}}} & \boldsymbol{K}_{ss} \end{bmatrix} \begin{bmatrix} \Delta \boldsymbol{q} \\ \Delta \boldsymbol{s} \end{bmatrix} = \begin{bmatrix} \boldsymbol{R}_{\varphi^{\mathrm{e}}} \\ \boldsymbol{R}_{s} \end{bmatrix}, \tag{4.49}$$

where $\Delta \boldsymbol{q} = [\Delta \boldsymbol{q}_1, \ldots, \Delta \boldsymbol{q}_{|\mathcal{I}|}]$ and $\Delta \boldsymbol{s} = [\Delta s_1, \ldots, \Delta s_{|\mathcal{I}|}]$ represent the incremental variation of the control variables for the deformed geometry and the phase-field parameter at time t_{n+1}. Moreover, $\boldsymbol{R}_{\varphi^{\mathrm{e}}}$ and $\boldsymbol{R}_s$ are given by (4.46) and (4.47), respectively, whereas $\boldsymbol{K}_{\varphi^{\mathrm{e}}\varphi^{\mathrm{e}}}$, $\boldsymbol{K}_{\varphi^{\mathrm{e}}s}$, $\boldsymbol{K}_{s\varphi^{\mathrm{e}}}$ and $\boldsymbol{K}_{ss}$ are tangential components. Note that we can solve the system either monolithically or sequentially by applying a staggered solution scheme, cf. chapter 3.

4.5 Numerical examples

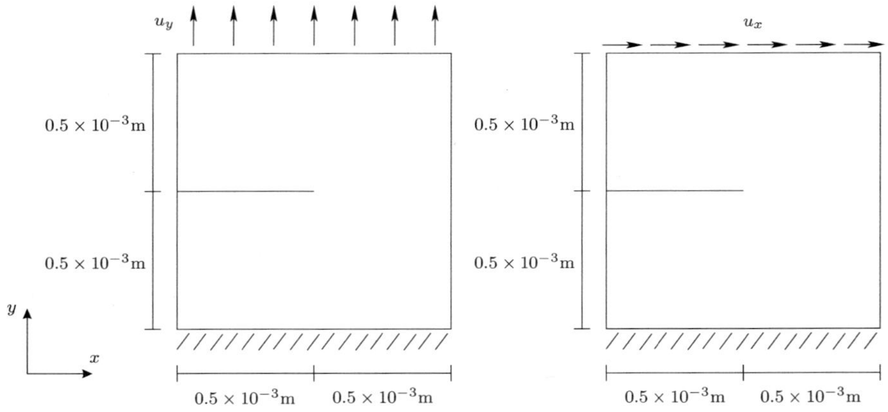

Figure 4.3: 2D crack test: Geometry and boundary conditions for the tension problem (left) and the shear problem (right).

The objective of this section is to present a series of numerical examples and to demonstrate the accuracy and performance of our proposed non-linear phase-field formulation in comparison to previously established methods. In particular, various quasi-static two dimensional and three dimensional benchmark problems are considered.

Throughout this section, we use either linear Lagrangian shape functions or quadratic B-Spline based shape functions whenever necessary to ensure global C^1-continuity for the higher order phase-field approach. Moreover, for the considered quasi-static examples considered in this section we make use of the more simple implicit Euler time integration scheme, whereas the proposed midpoint evaluation is applied for a transient impact problem within chapter 5.

4.5.1 2D crack test

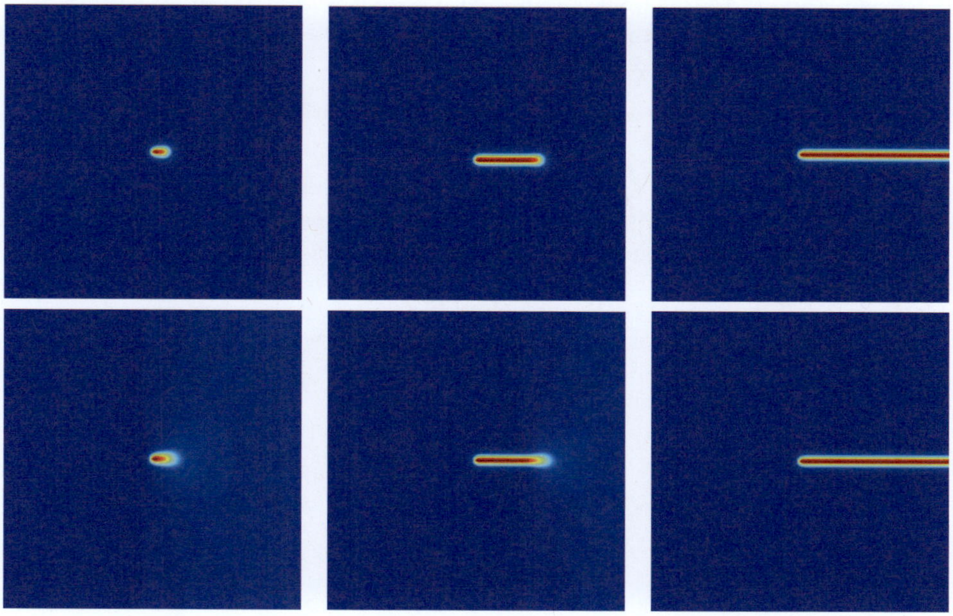

Figure 4.4: 2D tension test: Phase-field results for displacements of $u_y = [4.73,\ 4.83,\ 4.98] \times 10^{-6}$m using the cubic degradation function with $a_g = 0.1$ (upper row) and for displacements of $u_y = [4.47,\ 4.69,\ 4.83] \times 10^{-6}$ m using the quadratic degradation function (lower row). Results for the Neo-Hookean model with anisotropic split of the invariants are presented $s = 0$ (blue) and $s = 1$ (red).

The following two dimensional benchmark examples have been presented in Hesch & Weinberg [45] (see also Miehe et al. [64] for further details). The objective of these examples is to address the following points:

I) Comparison of the solution obtained by using linear Lagrangian shape functions to that obtained by using quadratic B-spline based shape functions for the approximation of the displacement and phase-field.

II) Comparison of the second order Allen-Cahn crack density functional (4.16) and the fourth order Cahn-Hilliard crack density functional (4.15) using quadratic B-spline basis functions in both cases. Note that the second order Allen-Cahn type approach will in general produce different results as the fourth order Cahn-Hilliard type approach, since the predefined length scale parameter controls the regularization zone differently in both approaches. However, we expect at least results which do not differ significantly and we want to outline the differences.

III) Study of the performance of the cubic degradation function as provided in (4.24) examining the solution of the problem.

IV) Comparison of the solution obtained with the proposed anisotropic split based on principal invariants with that obtained based on the principal eigenvalues of the deformation proposed in Hesch & Weinberg [45].

V) Study of the effect of mesh refinement upon the solution.

In order to investigate each of the aforementioned points, a fixed value of the length parameter $l = 7.5 \times 10^{-6}$m (interpreted as a material parameter) and a critical energy release rate $g_c = 2700\text{J/m}^2$ will be chosen. Moreover, the material behavior is assumed to be governed by a Neo-Hookean material model as defined in (4.27) with Lamé parameters $\mu = 80.769$GPa and $\lambda = 121.154$GPa, which correspond to a Young's modulus of $E = 210$GPa and a Poisson's ratio of $\nu = 0.3$ (steel-like material).

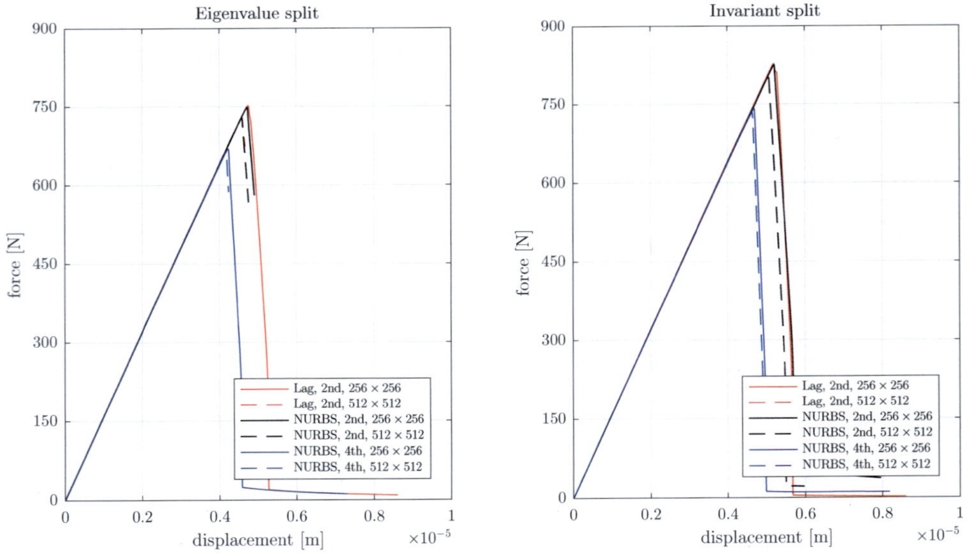

Figure 4.5: 2D tension test: Load-deflection curves for the tension problem using the eigenvalue and the invariant approach. Results for the second order (2nd) and the fourth order (4th) approach are compared using Lagrangian shape functions (Lag) and B-spline based shape functions (NURBS).

Tension test

A well-established benchmark problem is that of a symmetric tension test consisting of a squared plate with a horizontal notch (area with initialized fracture) with the geometry depicted in Figure 4.3 (left). The lower boundary is clamped as depicted, whereas the upper boundary is incrementally moved in (vertical) y-direction. The incremental displacement is given by $\Delta u_y = 10^{-8}$m until a displacement of $u_y = 4 \times 10^{-6}$m is reached, afterwards the incremental displacement is reduced by a factor of ten. A staggered solution scheme is used, such that we can control the error by the application of different incremental displacement step sizes. This is necessary to ensure a stable solution, since a monolithic scheme for this tension problem would require an arc length method to calculate the numerically unstable breakthrough of the crack. The uniform mesh consists of 256×256 and 512×512 elements, respectively, such that we obtain an element size of $h = 3.9 \times 10^{-6}$m and $h = 2 \times 10^{-6}$m, i.e. $h \approx l/2$ and $h \approx l/4$.

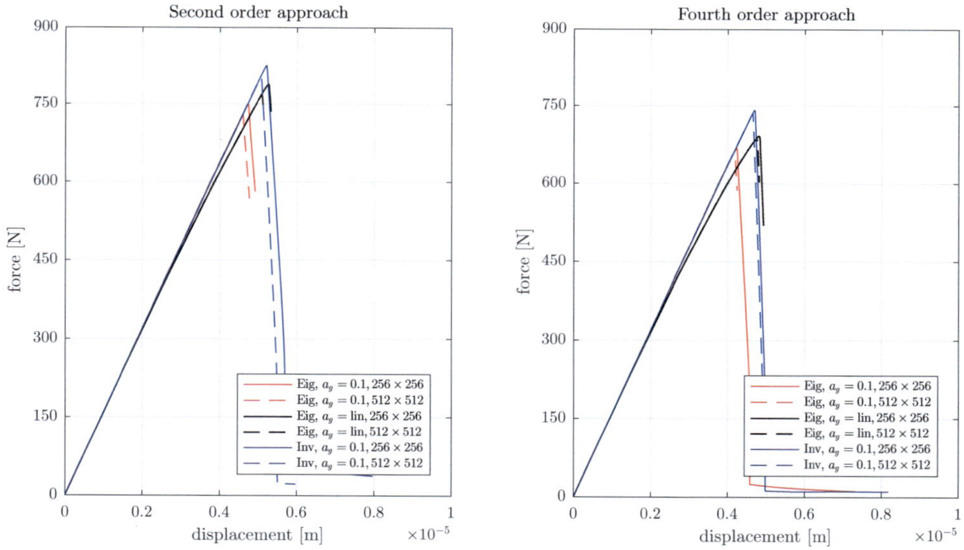

Figure 4.6: 2D tension test: Load-deflection curves for the tension problem for the second and the fourth order regularization profile. Results for the eigenvalue (Eig) and the invariant (Inv) approach are compared using the cubic degradation function with $a_\text{g} = 0.1$ and the linear degradation function ($a_\text{g} = \text{lin}$).

Snapshots of the phase-field result at different states of fracture are given in Figure 4.4, where the proposed anisotropic split based on the principal invariants is used.

Figure 4.5 shows a comparison of the load-deflection result provided by the anisotropic split based on principal eigenvalues (left) and that based on principal invariants (right). For each approach displayed in Figure 4.5, it is possible to observe that the solution obtained for both meshes, namely 256×256 and 512×512 elements are nearly identical, which indicates that for the coarser mesh the solution is already converged. Note that this addresses point V) above. Regarding point I), these two Figures show that for both anisotropic splits, the results obtained by using linear Lagrangian shape functions match exactly those obtained by using quadratic B-spline based shape functions when the second order Allen-Cahn crack density functional (4.16) is employed. Notice that this is expected for a sufficiently fine mesh with a converged solution.

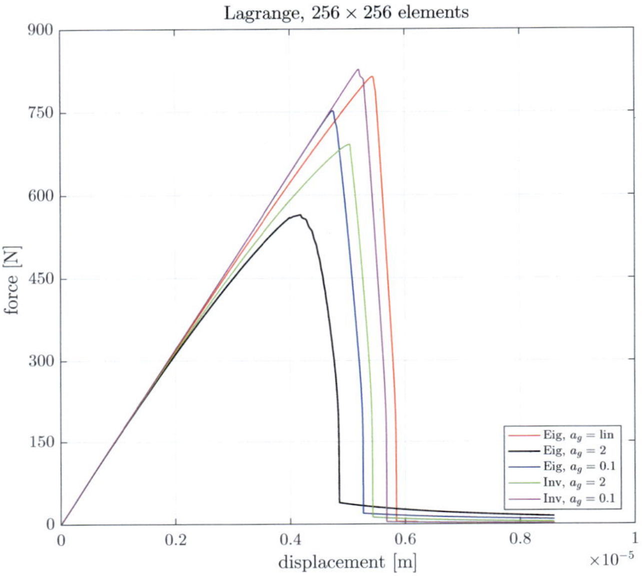

Figure 4.7: 2D tension test: Load-deflection curves for the tension problem. Study of different a_g parameters of the degradation function for the eigenvalue (Eig) and the invariant (Inv) approach.

Concerning point IV), we observe that the solution provided by both anisotropic splits (for the same choice of approximation spaces, degradation function and crack density functional) differs noticeably regarding the initialization of the crack. This is due to the fact, that different formulations of the elastic strain energy density functions are applied. Regarding point III), it is possible to observe the different influence of the cubic degradation function for a choice of the parameter $a_{\mathrm{g}} = 0.1$. Regarding point II), the results obtained when the fourth order Cahn-Hilliard crack density functional in (4.15) is used (in conjunction with quadratic B-splines) are slightly different from those obtained with the second order Allen-Cahn functional. Note that this was expected, as the fourth order functional regularizes the transition zone differently, i.e. the same length scale parameter l yields a different constitutive behavior in the material.

In Figure 4.6, a more extensive comparison of the results of the second order Allen-Cahn crack functional and those obtained with the fourth order Cahn-Hilliard functional is presented. For each choice of crack density functional, second order (Figure 4.6 (left)) and fourth order (Figure 4.6 (right)), both anisotropic splits have been employed using the cubic degradation function with a choice of $a_g = 0.1$. For completeness, a linear degradation function $(1-s)$ is also considered for the anisotropic split based on eigenvalues[1]. Notice that this linear degradation function is used in Hesch & Weinberg [45]. It can be observed that the cubic degradation function seems to be numerically more stable.

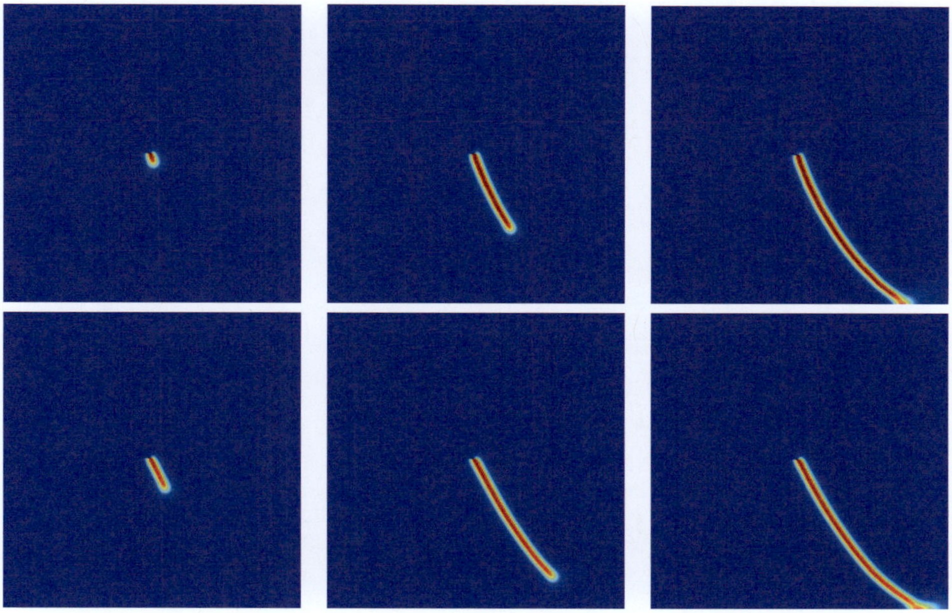

Figure 4.8: Shear problem: Phase-field results for displacements of $u_x = [8.5,\ 10.5,\ 12.2] \times 10^{-6}$m using the Neo-Hookean model with an anisotropic split of the principal invariants (upper row) and for displacements of $u_x = [8.5,\ 10.5,\ 11.05] \times 10^{-6}$m using the Neo-Hookean model with an anisotropic split of the principal eigenvalues (lower row). In both a parameter of $a_g = 0.1$ is used for the cubic degradation function.

This numerical stability is dramatically affected for the particular choice of $a_g = 2$ (degenerating into a quadratic degradation function) as observed in Figure 4.7, where a mesh of 256×256 elements and linear Lagrangian shape functions is employed for all the simulations. In contrast to the cubic degeneration function, the choice of the quadratic degeneration function leads to a softening before crack initialization for both anisotropic splits. Note that several curves for the eigenvalue split stop before a full breakthrough is reached. Since we do not use a trust-region or any other similar numerical optimization strategies, the additional material instability of this approach becomes apparent in this case. In contrast, the new approach based on principal invariants is numerically more stable.

[1] The linear degeneration function is considered within the exponent of the principal stretches.

Shear test

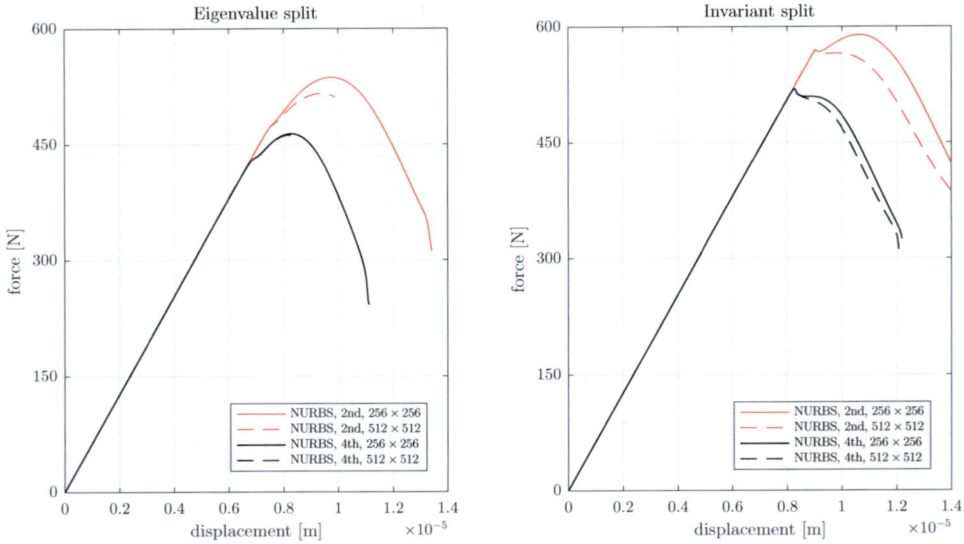

Figure 4.9: 2D shear test: Load-deflection curves for the shear problem using the eigenvalue and the invariant approach. Results for the second order (2nd) and the fourth order (4th) approach are compared using Lagrangian shape functions (Lag) and B-spline based shape functions (NURBS).

The geometry and boundary conditions of the shear test example are depicted in Figure 4.3 (right). The difference with respect to the previous example (tension test) relies on the application of a displacement in (horizontal) x-direction on the upper boundary of the body, yielding a shear type deformation. In contrast to the previous example, a uniform incremental displacement of $\Delta u_x = 10^{-8}$m is now applied on this boundary. The material parameters are chosen equally as for the tension test.

Again, a series of snapshots of the phase-field fracture result is presented in Figure 4.8 for both anisotropic splits.

The load-deflection curves for the two anisotropic splits are presented in Figure 4.9, yielding analogous conclusions to the tension test. As can be observed, the results for the mesh using 256×256 elements are nearly identical compared with the results of the mesh using 512×512 elements. Additionally, the newly proposed anisotropic split based on the principal invariants initializes slightly differently with respect to the anisotropic split based on the principal eigenvalues. Eventually, Figure 4.10 shows the load-deflection curves for the second and fourth order crack density functionals yielding similar conclusions to those obtained for the tension test in Figure 4.6.

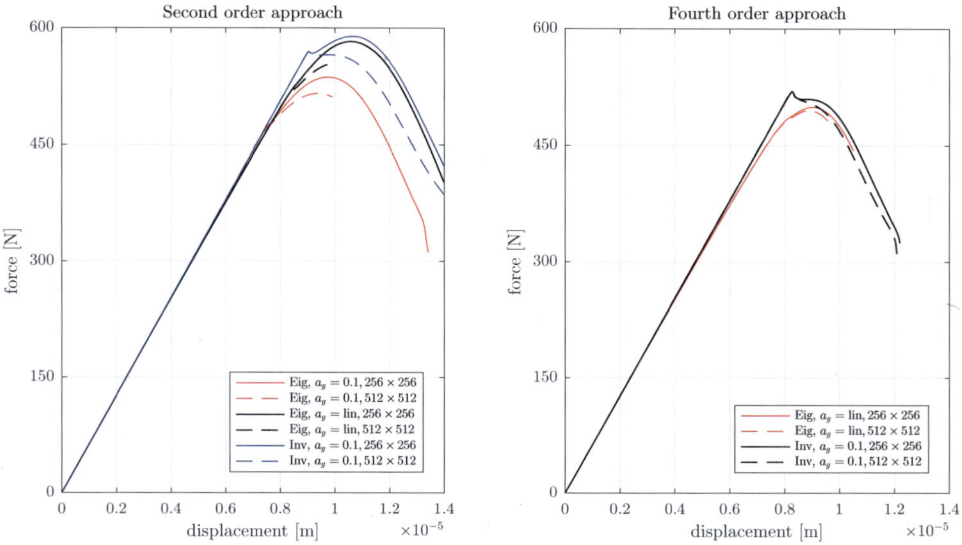

Figure 4.10: 2D shear test: Load-deflection curves for the shear problem for the second and the fourth order regularization profile. Results for the eigenvalue (Eig) and the invariant (Inv) approach are compared using the cubic degradation function with $a_{\mathrm{g}} = 0.1$ and the linear degradation function ($a_{\mathrm{g}} = \mathrm{lin}$).

4.5.2 2D large deformation test

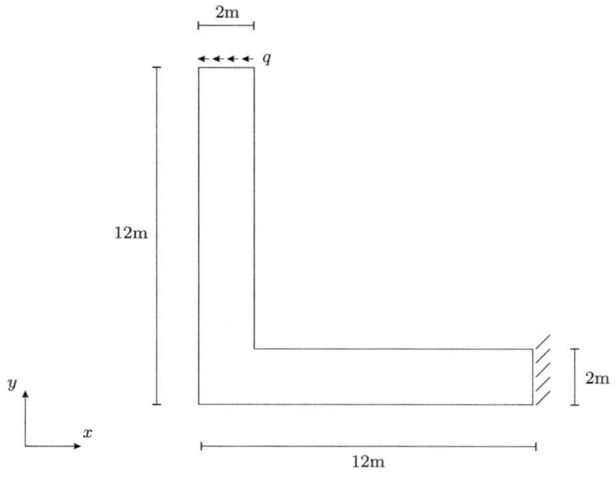

Figure 4.11: 2D large deformation test: Reference configuration and the applied boundary conditions.

The objective of this example is to illustrate the applicability of the formulation to scenarios where large deformations are obtained in the entire computational domain. With that in mind a Neo-Hookean constitutive model is considered with Lamé parameters $\mu = 0.071\text{GPa}$ and $\lambda = 0.286\text{GPa}$. This material setting corresponds to a synthetic substance with Young's modulus of $E = 0.2\text{GPa}$ and a Poisson ratio of $\nu = 0.4$. The critical energy release rate is chosen as $g_c = 150\text{kJ/m}^2$ in order to enable large deformations before crack initialization. Moreover, the length scale parameter reads $l = 0.01\text{m}$ and a value of $a_g = 0.1$ is used for the degradation function.

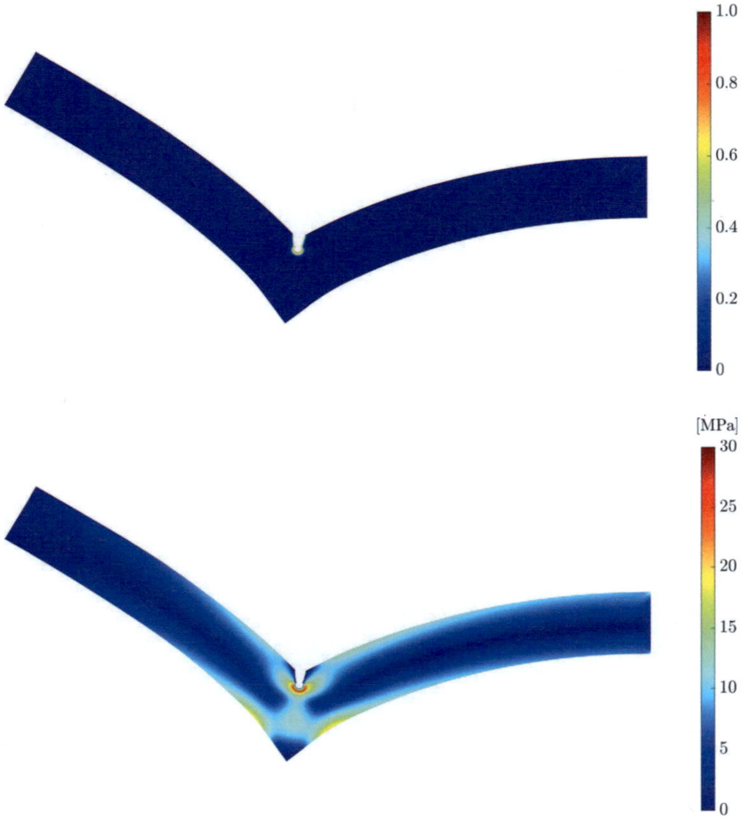

Figure 4.12: 2D large deformation test: Phase-field result (top) and von Mises stress distribution (bottom) after crack initialization at time $t = 0.15$. Broken elements are removed.

The reference configuration and the applied boundary conditions are given in Figure 4.11. The computational mesh consists of 158400 quadratic B-spline elements with in total 396500 degrees of freedom. A force per unit undeformed area of $\bar{\boldsymbol{T}} = [-q \times t, 0]$ with $q = 5 \times 10^6\text{N/m}^2$ is applied to the upper surface, controlled by a quasi time step size of $\Delta t = 0.001$ for each load increment. In addition, the L-shaped domain is completely constrained on the right hand side boundary.

Figure 4.12 displays the phase-field fracture result as well as the von Mises stress field after crack initialization at time $t = 0.15$. Note that all the broken elements have been removed for post-processing purposes.

4.5.3 3D crack test

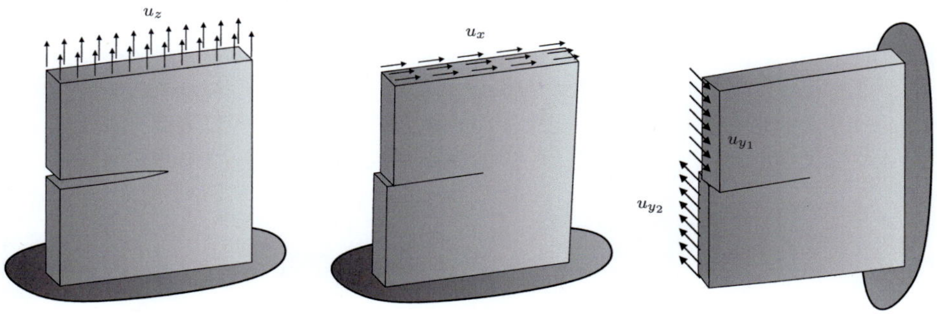

Figure 4.13: 3D crack test: Boundary conditions for mode I (left), mode II (middle) and mode III (right) crack propagation.

This section is to demonstrate the applicability of the newly defined phase-field approach using an anisotropic decomposition of the principal invariants in the context of three dimensional crack evolution. In particular, we consider a horizontal notched plate of size 1m × 0.2m × 1m. Taking the classical fracture modes I, II and III into account, the three dimensional plate is investigated within an opening, an in-plane shear and an out-of-plane shear displacement driven setting as depicted in Figure 4.13, see also Hesch et al. [44].

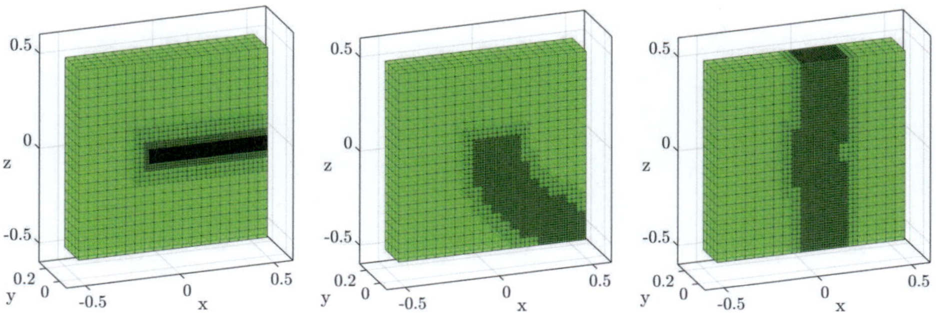

Figure 4.14: 3D crack test: Initial meshes of mode I (left), mode II (middle) and mode III (right)

The mesh used for each fracture mode consists of $20 \times 1 \times 20$ quadratic B-spline elements on level zero. Additionally, the finite element mesh is locally refined to capture the different crack path according to the specific fracture mode:

I) Three local refinement levels are used for the pure tension test, see Figure 4.14 (left). In total 46288 elements with overall 275354 degrees of freedom are employed. The displacement increment is set to $\Delta u_z = 5 \times 10^{-6}$m.

II) Two local refinement levels are used for the in-plane shear test, see Figure 4.14 (middle). In total 21900 elements with overall 102807 degrees of freedom are employed. The displacement increment is set to $\Delta u_x = 5 \times 10^{-6}$m.

III) Two local refinement levels are used for the out-of-plane shear test, see Figure 4.14 (right). In total 29040 elements with overall 135732 degrees of freedom are employed. The displacement increment is set to $\Delta u_{y_1} = \Delta u_{y_2} = 10 \times 10^{-6}$m.

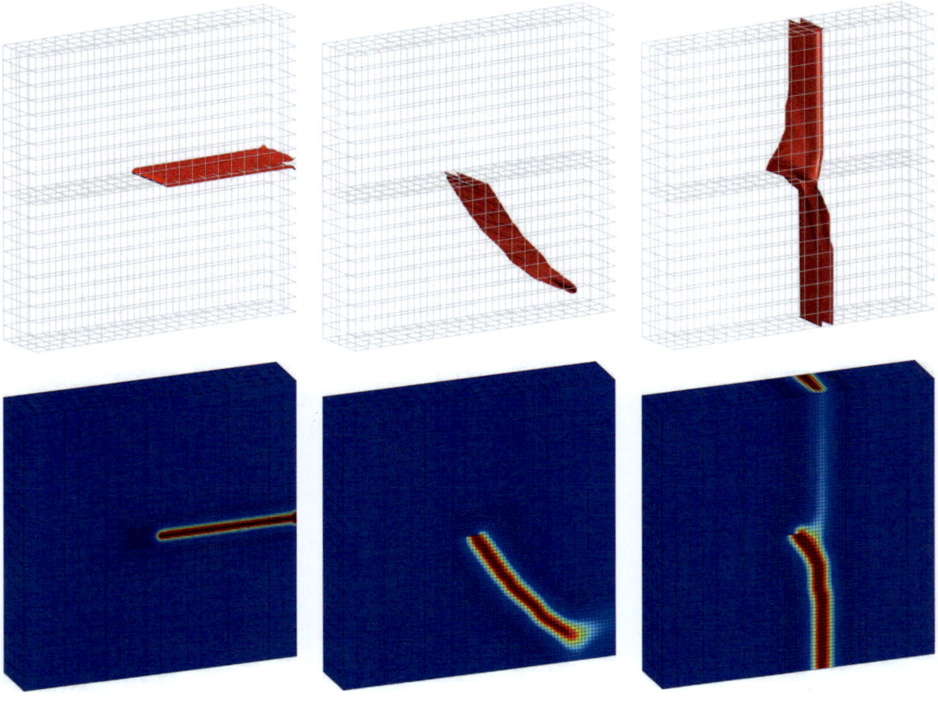

Figure 4.15: 3D crack test: Final phase-field results of mode I (left, for displacement $u_z = 0.26 \times 10^{-3}$m), mode II (middle, for displacement $u_x = 0.37 \times 10^{-3}$m) and mode III (right, for displacement $u_{y_1} = u_{y_2} = 1.74 \times 10^{-3}$m)

The constitutive behavior is assumed to be governed by a Mooney-Rivlin material law where the parameters correspond to a steel-like material and take the values $\alpha = 26.923\text{GPa}$, $\beta = 13.462\text{GPa}$ and $\kappa = 175\text{GPa}$. In addition, the phase-field parameters are set to $g_c = 2700\text{J/m}^2$, $l = 0.0081\text{m}$ and $a_g = 0.085$ for mode I, whereas $l = 0.0138\text{m}$ and $a_g = 0.05$ are used for mode II and III.

The phase-field is graphically presented using isosurfaces for $s = 0.8$ in Figure 4.15, displaying detailed snapshots of the crack progression for the different modes. Note that we constructed the B-spline meshes using a higher order domain decomposition between two rectangular blocks starting at the center up to the right hand boundary, such that we obtain an initial crack from the left hand boundary to the center. The meshes in Figure 4.15 indicate this construction by displaying the internal interface. Figure 4.16 shows the corresponding von Mises stress distribution for the three modes at certain time steps.

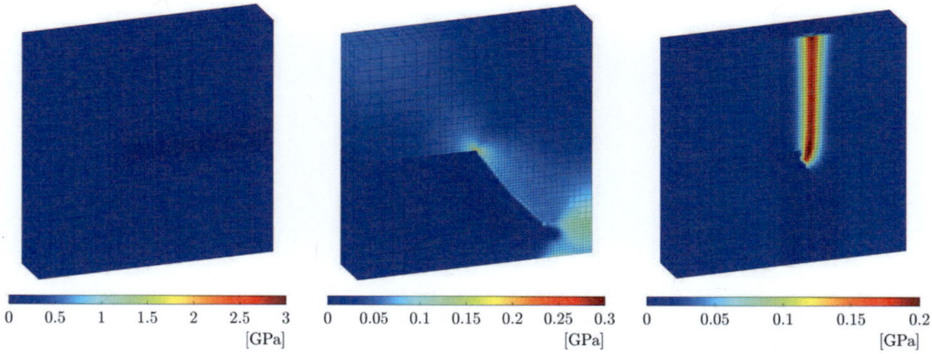

Figure 4.16: 3D crack test: Von Mises stress results of mode I (left, for displacement $u_z = 0.26 \times 10^{-3}\text{m}$), mode II (middle, for displacement $u_x = 0.37 \times 10^{-3}\text{m}$) and mode III (right, for displacement $u_{y_1} = u_{y_2} = 1.74 \times 10^{-3}\text{m}$)

To summarize, the proposed fourth order model using the cubic degradation function and the proposed polyconvex anisotropic split of the principal invariants allows for the prediction of complex three dimensional crack propagation. Especially the mode III shear test results in very complex, three dimensional fracture patterns.

5 Contact mechanics

In this chapter we consider thermomechanical as well as fracture mechanical contact and impact problems within the context of IGA. Based on the first and second laws, the thermodynamic foundations of frictional interfaces govern the formulation of appropriate constitutive laws and additionally play a major role in the construction of the accompanying numerical schemes. Here, the emphasis will be on the numerical aspects. In particular, we aim at a variationally consistent mortar approach for B-spline and NURBS discretized and hierarchically refined bodies in contact. To be specific, we employ a quadratic B-spline or NURBS discretization with local refinements for the contact interface, whereas linear Lagrangian shape functions are used for the discrete traction field. This approach leads to a simplified and fast mortar framework, which has been proposed for domain decomposition problems in Hesch & Betsch [41], for thermomechanical contact problems in Dittmann et al. [27] and for fracture mechanical contact problems in Hesch et al. [42]. The mortar projections will be calculated via a newly developed segmentation procedure of the surface intersections. For the thermal contributions, the mortar concept will be applied by introducing so-called triple mortar integrals to accurately capture the frictional dissipation contribution to the contact heat flux and to establish a correct thermal interaction among the contacting surfaces. Eventually, numerical examples are investigated for both, thermomechanical as well as fracture mechanical contact problems. These examples will demonstrate the capabilities of the entire framework presented in this work.

5.1 Contact formulation

Assuming two thermoelastic bodies $i \in \{1, 2\}$ in a contact situation, the boundaries are further subdivided and satisfy

$$\partial\mathcal{B}_0^{(i),\mathrm{c}} \cup \partial\mathcal{B}_0^{(i),\varphi} \cup \partial\mathcal{B}_0^{(i),T} = \partial\mathcal{B}_0^{(i),\mathrm{c}} \cup \partial\mathcal{B}_0^{(i),\theta} \cup \partial\mathcal{B}_0^{(i),Q} = \partial\mathcal{B}_0^{(i)} \tag{5.1}$$

and

$$\begin{aligned} \partial\mathcal{B}_0^{(i),\mathrm{c}} \cap \partial\mathcal{B}_0^{(i),\varphi} = \partial\mathcal{B}_0^{(i),\varphi} \cap \partial\mathcal{B}_0^{(i),T} = \partial\mathcal{B}_0^{(i),T} \cap \partial\mathcal{B}_0^{(i),\mathrm{c}} = \emptyset, \\ \partial\mathcal{B}_0^{(i),\mathrm{c}} \cap \partial\mathcal{B}_0^{(i),\theta} = \partial\mathcal{B}_0^{(i),\theta} \cap \partial\mathcal{B}_0^{(i),Q} = \partial\mathcal{B}_0^{(i),Q} \cap \partial\mathcal{B}_0^{(i),\mathrm{c}} = \emptyset. \end{aligned} \tag{5.2}$$

In this connection, $\partial\mathcal{B}_0^{(i),\mathrm{c}}$ represents the potential contact area on surface (i).

Assuming that the local linear momentum balance equation across the contact interface

$$\boldsymbol{t}^{(1)}\mathrm{d}A = -\boldsymbol{t}^{(2)}\mathrm{d}A, \tag{5.3}$$

where $\boldsymbol{t}^{(i)}$ denotes the Piola traction associated with surface $\partial\mathcal{B}_0^{(i),\mathrm{c}}$, is valid, we state the global energy balance of the interface as follows

$$\int_{\partial\mathcal{B}_0^{(1),\mathrm{c}}} \dot{e}_\mathrm{c} \mathrm{d}A = \int_{\partial\mathcal{B}_0^{(1),\mathrm{c}}} (-\boldsymbol{t}^{(1)} \cdot (\dot{\boldsymbol{\varphi}}^{(1)} - \dot{\boldsymbol{\varphi}}^{(2)}) + Q_\mathrm{c}^{(1)} + Q_\mathrm{c}^{(2)}) \mathrm{d}A. \tag{5.4}$$

Here, e_c is the stored inner energy density measured per unit area on the contact interface, $(\dot{\boldsymbol{\varphi}}^{(1)} - \dot{\boldsymbol{\varphi}}^{(2)})$ describes the relative velocity between both contact surfaces and $Q_\mathrm{c}^{(i)}$ denotes the heat supply across the respective surface $\partial\mathcal{B}_0^{(i),\mathrm{c}}$ on the contact interface. Note that the heat supplies are not required to be equal and opposite, such that a heat generation on the contact interface is possible. Next, we apply the localization theorem and obtain the local form of the balance equation as

$$\dot{e}_\mathrm{c} = -\boldsymbol{t}^{(1)} \cdot (\dot{\boldsymbol{\varphi}}^{(1)} - \dot{\boldsymbol{\varphi}}^{(2)}) + Q_\mathrm{c}^{(1)} + Q_\mathrm{c}^{(2)}, \tag{5.5}$$

which can be recast in terms of the local Helmholtz boundary energy density $\Psi_\mathrm{c} = e_\mathrm{c} - \theta_\mathrm{c}\eta_\mathrm{c}$ as

$$\dot{\Psi}_\mathrm{c} + \theta_\mathrm{c}\dot{\eta}_\mathrm{c} + \eta_\mathrm{c}\dot{\theta}_\mathrm{c} = -\boldsymbol{t}^{(1)} \cdot (\dot{\boldsymbol{\varphi}}^{(1)} - \dot{\boldsymbol{\varphi}}^{(2)}) + Q_\mathrm{c}^{(1)} + Q_\mathrm{c}^{(2)}. \tag{5.6}$$

Therein, η_c and θ_c are the entropy density per unit area on the contact interface and the absolute temperature of the interface, respectively.

Similar to (3.13), we postulate that the total entropy production on contact interface is greater than or equal to zero, i.e. in local form the second law of thermodynamics for the interface becomes

$$\dot{\eta}_\mathrm{c} \geq \frac{Q_\mathrm{c}^{(1)}}{\theta^{(1)}} + \frac{Q_\mathrm{c}^{(2)}}{\theta^{(2)}}, \tag{5.7}$$

where $\theta^{(i)}$ is the temperature of the body (i) as the contact interface is approached. Combining (5.6) and (5.7), we obtain the inequality

$$-\dot{\Psi}_\mathrm{c} - \eta_\mathrm{c}\dot{\theta}_\mathrm{c} - \boldsymbol{t}^{(1)} \cdot (\dot{\boldsymbol{\varphi}}^{(1)} - \dot{\boldsymbol{\varphi}}^{(2)}) + Q_\mathrm{c}^{(1)} \left(1 - \frac{\theta_c}{\theta^{(1)}}\right) + Q_\mathrm{c}^{(2)} \left(1 - \frac{\theta_c}{\theta^{(2)}}\right) \geq 0. \tag{5.8}$$

To account for different contact reactions in normal and tangential direction, we may decompose the Piola traction into normal and frictional tractions as follows

$$\boldsymbol{t}^{(1)} = -t_\mathrm{N}\boldsymbol{n} - \boldsymbol{t}_\mathrm{T} \qquad \text{with} \qquad \boldsymbol{t}_\mathrm{T} \cdot \boldsymbol{n} = 0, \tag{5.9}$$

where $\boldsymbol{n} = \boldsymbol{n}^{(2)}$ denotes the current outward normal vector on $\partial\mathcal{B}_0^{(2),\mathrm{c}}$. Accordingly, we define the local normal and tangential gap functions as

$$g_\mathrm{N} = \boldsymbol{n} \cdot (\boldsymbol{\varphi}^{(1)} - \boldsymbol{\varphi}^{(2)}) \qquad \text{and} \qquad \boldsymbol{g}_\mathrm{T} = (\boldsymbol{I} - \boldsymbol{n} \otimes \boldsymbol{n}) \cdot (\boldsymbol{\varphi}^{(1)} - \boldsymbol{\varphi}^{(2)}), \tag{5.10}$$

cf. Puso & Laursen [77]. Note that the components of the tangential traction and the tangential gap function can be written in terms of co-/contravariant basis vectors

$$\boldsymbol{t}_\mathrm{T} = t_{\mathrm{T},\alpha}\boldsymbol{a}^\alpha \qquad \text{and} \qquad \boldsymbol{g}_\mathrm{T} = g_\mathrm{T}^\alpha \boldsymbol{a}_\alpha, \tag{5.11}$$

which is often useful for the definition of the constitutive frictional law (e.g. in case of anisotropic friction).

5.2 Constitutive theory for contact mechanics

The tangential gap function can be further decomposed additively into a reversible (elastic) part $\boldsymbol{g}_{\mathrm{T}}^{\mathrm{e}}$ and an irreversible (inelastic) part $\boldsymbol{g}_{\mathrm{T}}^{\mathrm{s}}$. Thus, we define the free Helmholtz boundary energy as a function dependent on the normal penetration, the elastic part of the tangential slip and the interface temperature

$$\Psi_{\mathrm{c}} = \Psi_{\mathrm{c}}(g_{\mathrm{N}}, \boldsymbol{g}_{\mathrm{T}}^{\mathrm{e}}, \theta_{\mathrm{c}}). \tag{5.12}$$

Now, differentiating with respect to time and insertion into (5.8) yields

$$\begin{aligned} \left(t_{\mathrm{N}} - \frac{\partial \Psi_{\mathrm{c}}}{\partial g_{\mathrm{N}}}\right) \dot{g}_{\mathrm{N}} + \boldsymbol{t}_{\mathrm{T}} \cdot \dot{\boldsymbol{g}}_{\mathrm{T}} - \frac{\partial \Psi_{\mathrm{c}}}{\partial \boldsymbol{g}_{\mathrm{T}}^{\mathrm{e}}} \cdot \dot{\boldsymbol{g}}_{\mathrm{T}}^{\mathrm{e}} - \left(\eta_{\mathrm{c}} + \frac{\partial \Psi_{\mathrm{c}}}{\partial \theta_{\mathrm{c}}}\right) \dot{\theta}_{\mathrm{c}} \\ + Q_{\mathrm{c}}^{(1)} \left(1 - \frac{\theta_c}{\theta^{(1)}}\right) + Q_{\mathrm{c}}^{(2)} \left(1 - \frac{\theta_c}{\theta^{(2)}}\right) \geq 0. \end{aligned} \tag{5.13}$$

This inequality has to be satisfied for each admissible evolution of the system. Assuming that the rates of change $\dot{g}_{\mathrm{N}}$, $\dot{\boldsymbol{g}}_{\mathrm{T}}^{\mathrm{e}}$ and $\dot{\theta}_{\mathrm{c}}$ can be arbitrary, we obtain the following constitutive relations

$$t_{\mathrm{N}} = \frac{\partial \Psi_{\mathrm{c}}}{\partial g_{\mathrm{N}}}, \qquad \boldsymbol{t}_{\mathrm{T}} = \frac{\partial \Psi_{\mathrm{c}}}{\partial \boldsymbol{g}_{\mathrm{T}}^{\mathrm{e}}} \qquad \text{and} \qquad \eta_{\mathrm{c}} = -\frac{\partial \Psi_{\mathrm{c}}}{\partial \theta_{\mathrm{c}}}, \tag{5.14}$$

along with the dissipation inequality

$$\boldsymbol{t}_{\mathrm{T}} \cdot \dot{\boldsymbol{g}}_{\mathrm{T}}^{\mathrm{s}} + Q_{\mathrm{c}}^{(1)} \left(1 - \frac{\theta_c}{\theta^{(1)}}\right) + Q_{\mathrm{c}}^{(2)} \left(1 - \frac{\theta_c}{\theta^{(2)}}\right) \geq 0. \tag{5.15}$$

In order to satisfy (5.15), we postulate the existence of a convex, but in general non-differentiable and infinite-valued dissipation potential

$$D_{\mathrm{c}} = D_{\mathrm{c}}(\dot{\boldsymbol{g}}_{\mathrm{T}}^{\mathrm{s}}, g_{\theta}^{(1)}, g_{\theta}^{(2)}), \tag{5.16}$$

where $g_{\theta}^{(i)} = \theta^{(i)} - \theta_{\mathrm{c}}$ denotes the respective temperature jump. Moreover, the thermodynamic forces correlate to the subdifferential of the dissipation potential, namely

$$\left(\boldsymbol{t}_{\mathrm{T}}, \frac{Q_{\mathrm{c}}^{(1)}}{\theta^{(1)}}, \frac{Q_{\mathrm{c}}^{(2)}}{\theta^{(2)}}\right) \in \partial D_{\mathrm{c}}(\dot{\boldsymbol{g}}_{\mathrm{T}}^{\mathrm{s}}, g_{\theta}^{(1)}, g_{\theta}^{(2)}), \tag{5.17}$$

see also Moreau [67], Johansson & Klarbring [53] and Oancea & Laursen [69].

5.2.1 Free Helmholtz energy and dissipation potential

Now, we consider more specific constitutive relationships for the contact interface satisfying the above conditions. In particular, we specify the free Helmholtz boundary energy density function as

$$\Psi_{\mathrm{c}}(g_{\mathrm{N}}, \boldsymbol{g}_{\mathrm{T}}^{\mathrm{e}}, \theta_{\mathrm{c}}) = \underbrace{I_{\mathbb{R}^+}(g_{\mathrm{N}}) + \frac{1}{2}\epsilon_{\mathrm{T}} \boldsymbol{g}_{\mathrm{T}}^{\mathrm{e}} \cdot \boldsymbol{g}_{\mathrm{T}}^{\mathrm{e}}}_{\Psi_{\mathrm{c}}^{\mathrm{mech}}} \underbrace{- \frac{c_{\mathrm{c}}}{2\theta_{\mathrm{ref}}}(\theta_{\mathrm{c}} - \theta_{\mathrm{ref}})^2}_{\Psi_{\mathrm{c}}^{\mathrm{the}}}, \tag{5.18}$$

where the indicator function

$$I_{\mathbb{R}^+}(g_{\rm N}) = \begin{cases} 0 & \text{if } g_{\rm N} \geq 0 \\ +\infty & \text{if } g_{\rm N} < 0 \end{cases} \tag{5.19}$$

within the mechanical part $\Psi_{\rm c}^{\rm mech}$ is associated with an infinite stiffness of the interface in normal direction, whereas the tangential penalty parameter $\epsilon_{\rm T} \geq 0$ can be considered as tangential stiffness and $c_{\rm c} \geq 0$ denotes the heat capacity measured per unit surface on the interface, see Laursen [56]. $\Psi_{\rm c}^{\rm the}$ denotes the thermal contribution which is associated with a decreases in the free Helmholtz energy due to stored heat on the interface. Moreover, an admissible dissipation function can be selected as

$$D_{\rm c}(\dot{\boldsymbol{g}}_{\rm T}^{\rm s}, g_\theta^{(1)}, g_\theta^{(2)}) = \underbrace{|t_{\rm N}|\mu_{\rm c}\|\dot{\boldsymbol{g}}_{\rm T}^{\rm s}\|}_{D_{\rm c}^{\rm fric}} + \underbrace{\frac{1}{2}\frac{k_{\rm c}^{(1)}|t_{\rm N}|}{\theta^{(1)}}(g_\theta^{(1)})^2 + \frac{1}{2}\frac{k_{\rm c}^{(2)}|t_{\rm N}|}{\theta^{(2)}}(g_\theta^{(2)})^2}_{D_{\rm c}^{\rm heat}} \geq 0, \tag{5.20}$$

where $D_{\rm c}^{\rm fric}$ denotes the frictional dissipation and $D_{\rm c}^{\rm heat}$ is the dissipation due to the irreversibility of heat transfer across the interface. Here, the friction coefficient $\mu_{\rm c} \geq 0$ is assumed to be constant and the heat transfer coefficient $k_{\rm c}^{(i)}|t_{\rm N}| \geq 0$ is modeled via the contact pressure.

5.2.2 Contact pressure

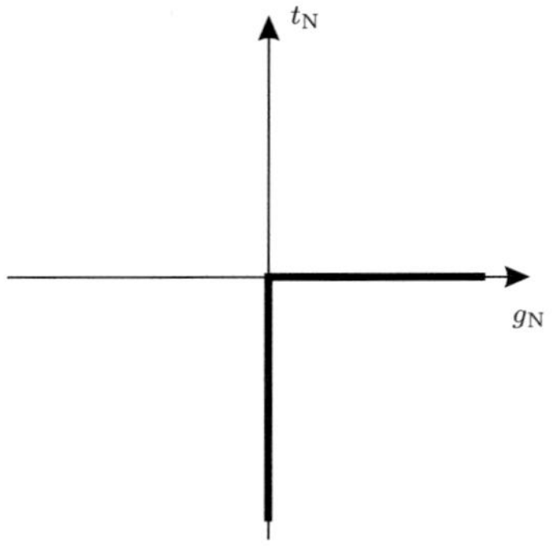

Figure 5.1: Illustration of the Karush-Kuhn-Tucker condition for non-penetration.

In order to obtain an expression for the contact pressure, we apply $(5.14)_1$ along with (5.18), which leads to the standard Karush-Kuhn-Tucker conditions for non-penetration in normal direction

$$\begin{aligned} g_{\rm N} &\geq 0, \\ t_{\rm N} &\leq 0, \\ t_{\rm N} g_{\rm N} &= 0, \end{aligned} \tag{5.21}$$

see Figure 5.1 for illustration. Note that the stored energy on the interface due to the normal traction is assumed to be equal to zero. Constitutive relations to account for, e.g. finite stiffness of the interface or adhesion effects, can be embedded in straightforward manner by modifying the first term in (5.18).

5.2.3 Friction law

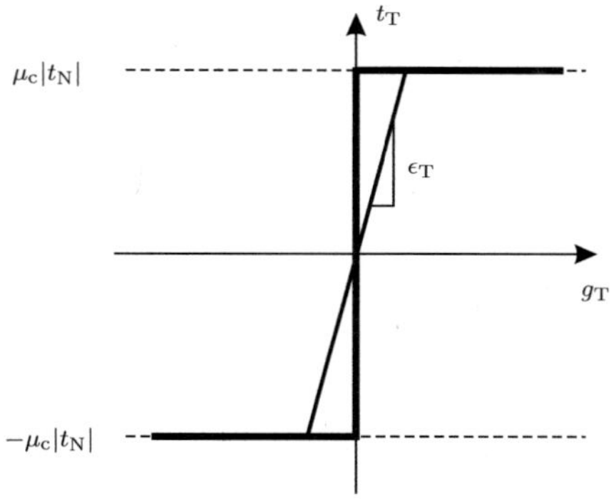

Figure 5.2: One dimensional illustration of Coulomb's friction law and penalization of the stick condition.

Combining $(5.14)_2$, (5.17), (5.18) and (5.20), we obtain the tangential traction as

$$\boldsymbol{t}_\mathrm{T} = \begin{cases} \epsilon_\mathrm{T} \boldsymbol{g}_\mathrm{T}^\mathrm{e} & \text{if } \|\dot{\boldsymbol{g}}_\mathrm{T}^\mathrm{s}\| = 0, \\ \mu_\mathrm{c}|t_\mathrm{N}| \frac{\dot{\boldsymbol{g}}_\mathrm{T}^\mathrm{s}}{\|\dot{\boldsymbol{g}}_\mathrm{T}^\mathrm{s}\|} & \text{if } \|\dot{\boldsymbol{g}}_\mathrm{T}^\mathrm{s}\| \neq 0, \end{cases} \tag{5.22}$$

see Figure 5.2 for illustration. Given the definitions above and following standard arguments as outlined, e.g. in Yang & Laursen [102] and in Puso & Laursen [77], Coulomb's friction law can be written as

$$\begin{aligned} \hat{\phi}_\mathrm{c} = \|\boldsymbol{t}_\mathrm{T}\| - \mu_\mathrm{c}|t_\mathrm{N}| &\leq 0, \\ \dot{\zeta} &\geq 0, \\ \hat{\phi}_\mathrm{c}\dot{\zeta} &= 0. \end{aligned} \tag{5.23}$$

Here, $\hat{\phi}_\mathrm{c}$ is denoted as slip function and $\dot{\zeta}$ is a consistency parameter, where $\dot{\zeta} = 0$ represents stick and $\dot{\zeta} > 0$ slip, in analogy to the plastic multiplier in plasticity. Now, the Lie derivative $\mathscr{L}\boldsymbol{t}_\mathrm{T} = \dot{t}_{\mathrm{T},\alpha}\boldsymbol{a}^\alpha$ of the frictional traction reads

$$\mathscr{L}\boldsymbol{t}_\mathrm{T} = \epsilon_\mathrm{T}\left(\dot{\boldsymbol{g}}_\mathrm{T} - \dot{\zeta}\frac{\boldsymbol{t}_\mathrm{T}}{\|\boldsymbol{t}_\mathrm{T}\|}\right). \tag{5.24}$$

Note that in contrast to the normal contributions, where we enforce a geometrical non-penetration condition via Lagrange multipliers, we enforce the stick condition in tangential direction via a penalty regularization. Other choices are possible and can be applied without substantially changing the proposed framework.

5.2.4 Thermal contact interface

Thermodynamically consistent formulations for the heat supplies can be deduced from (5.17) and (5.20) as

$$Q_{\mathrm{c}}^{(i)} = k_{\mathrm{c}}^{(i)} |t_{\mathrm{N}}| g_{\theta}^{(i)}. \tag{5.25}$$

Moreover, insertion of (5.14) into (5.6) yields

$$\frac{c_{\mathrm{c}}}{\theta_{\mathrm{ref}}} \dot{\theta}_{\mathrm{c}} \theta_{\mathrm{c}} = \boldsymbol{t}_{\mathrm{T}} \cdot \dot{\boldsymbol{g}}_{\mathrm{T}}^{\mathrm{s}} + Q_{\mathrm{c}}^{(1)} + Q_{\mathrm{c}}^{(2)}. \tag{5.26}$$

Now, assuming that the heat capacity of the interface consisting of, e.g. trapped wear debris or surface coating, is small enough to be neglected, we can eliminate the interface temperature and obtain explicit expressions for the heat fluxes across the interface as

$$Q_{\mathrm{c}}^{(1)} = \frac{k_{\mathrm{c}}^{(1)} k_{\mathrm{c}}^{(2)}}{k_{\mathrm{c}}^{(1)} + k_{\mathrm{c}}^{(2)}} |t_{\mathrm{N}}| (\theta^{(1)} - \theta^{(2)}) - \frac{k_{\mathrm{c}}^{(1)}}{k_{\mathrm{c}}^{(1)} + k_{\mathrm{c}}^{(2)}} \boldsymbol{t}_{\mathrm{T}} \cdot \dot{\boldsymbol{g}}_{\mathrm{T}}^{\mathrm{s}} \tag{5.27}$$

and

$$Q_{\mathrm{c}}^{(2)} = -\frac{k_{\mathrm{c}}^{(1)} k_{\mathrm{c}}^{(2)}}{k_{\mathrm{c}}^{(1)} + k_{\mathrm{c}}^{(2)}} |t_{\mathrm{N}}| (\theta^{(1)} - \theta^{(2)}) - \frac{k_{\mathrm{c}}^{(2)}}{k_{\mathrm{c}}^{(1)} + k_{\mathrm{c}}^{(2)}} \boldsymbol{t}_{\mathrm{T}} \cdot \dot{\boldsymbol{g}}_{\mathrm{T}}^{\mathrm{s}}. \tag{5.28}$$

The above formulations can be further simplified by introducing the following abbreviations

$$k_{\mathrm{c}} = \frac{k_{\mathrm{c}}^{(1)} k_{\mathrm{c}}^{(2)}}{k_{\mathrm{c}}^{(1)} + k_{\mathrm{c}}^{(2)}} \quad \text{and} \quad \gamma_{\mathrm{c}}^{(i)} = \frac{k_{\mathrm{c}}^{(i)}}{k_{\mathrm{c}}^{(1)} + k_{\mathrm{c}}^{(2)}}, \tag{5.29}$$

such that

$$Q_{\mathrm{c}}^{(1)} = k_{\mathrm{c}} |t_{\mathrm{N}}| g_{\theta} - \gamma_{\mathrm{c}}^{(1)} D_{\mathrm{c}}^{\mathrm{fric}} \quad \text{and} \quad Q_{\mathrm{c}}^{(2)} = -k_{\mathrm{c}} |t_{\mathrm{N}}| g_{\theta} - \gamma_{\mathrm{c}}^{(2)} D_{\mathrm{c}}^{\mathrm{fric}}, \tag{5.30}$$

where $g_{\theta} = \theta^{(1)} - \theta^{(2)}$ and $D_{\mathrm{c}}^{\mathrm{fric}} = \boldsymbol{t}_{\mathrm{T}} \cdot \dot{\boldsymbol{g}}_{\mathrm{T}}^{\mathrm{s}} = |t_{\mathrm{N}}| \mu_{\mathrm{c}} \|\dot{\boldsymbol{g}}_{\mathrm{T}}^{\mathrm{s}}\|$. The heat fluxes are composed of two contributions describing thermal interaction and frictional dissipation. Note that the constitutive modeling of the thermal interaction is based on the consideration of various heat exchange mechanisms across surfaces in vicinity, see, e.g. Madhusudana [59], Bahrami et al. [7] and Persson et al. [74]. In a continuum setting, primarily three mechanisms are of concern: conduction, convection and radiation. Among these, conduction is a contact interaction due to the temperature jump g_{θ} across the contact interface, whereas the latter two are of non-contact type. The radiation contribution is often negligible unless the temperatures involved are too high. Macroscopically contacting surfaces, on the other hand, may simultaneously involve the remaining two mechanisms due to inherent roughness since the microscopic interface topography not only has contacting asperities but also gaps containing an interstitial material which serves as a medium for convection. Moreover, in view of the multiscale nature of roughness, the thermal interaction of the contacting asperities is also governed by both mechanisms. Due to the small length scales associated with the gaps, the convective mechanism degenerates into conduction through the interstitial medium. Finally, an increasing contact pressure leads to higher microscopic conformity among the surfaces in contact such that the macroscopic interface resistance decreases. Consequently, the heat exchange can be modeled as a function of g_θ and t_N. In addition, the relative effusivities satisfy $\gamma^{(1)} + \gamma^{(2)} = 1$, such that the mechanical energy is directly dissipated into the thermal field without changing the inner energy, i.e.

$$Q_\mathrm{c}^{(1)} + Q_\mathrm{c}^{(2)} = -D_\mathrm{c}^\mathrm{fric}. \tag{5.31}$$

Following the arguments in Oancea & Laursen [69], stating that the elastic part of the tangential gap and the associated stored energy are small enough to be neglected, which correlates to a high value of the tangential penalty parameter, we obtain

$$\int\limits_{\partial\mathcal{B}_0^{(1),c}} \dot{e}_c \,\mathrm{d}A = \int\limits_{\partial\mathcal{B}_0^{(1),c}} (t_\mathrm{N}\dot{g}_\mathrm{N} + \boldsymbol{t}_\mathrm{T} \cdot \dot{\boldsymbol{g}}_\mathrm{T} + Q_\mathrm{c}^{(1)} + Q_\mathrm{c}^{(2)}) \,\mathrm{d}A = 0. \tag{5.32}$$

5.3 Variational formulation

Table 5.1: Variational formulation of the thermomechanical contact problem

1. Mechanical field

$$\begin{aligned}\sum_i \int\limits_{\mathcal{B}_0^{(i)}} (\delta\boldsymbol{\varphi}^{(i)} \cdot \dot{\boldsymbol{\pi}}^{(i)} + \boldsymbol{P}^{(i)} : \nabla(\delta\boldsymbol{\varphi}^{(i)}))\,\mathrm{d}V - \sum_i \int\limits_{\mathcal{B}_0^{(i)}} \delta\boldsymbol{\varphi}^{(i)} \cdot \boldsymbol{B}^{(i)} \,\mathrm{d}V \\ - \sum_i \int\limits_{\partial\mathcal{B}_0^{(i),T}} \delta\boldsymbol{\varphi}^{(i)} \cdot \bar{\boldsymbol{T}}^{(i)} \,\mathrm{d}A + \int\limits_{\partial\mathcal{B}_0^{(1),c}} (t_\mathrm{N}\delta g_\mathrm{N} + \boldsymbol{t}_\mathrm{T} \cdot \delta\boldsymbol{g}_\mathrm{T})\,\mathrm{d}A = 0\end{aligned} \tag{5.33}$$

2. Thermal field

$$\begin{aligned}\sum_i \int\limits_{\mathcal{B}_0^{(i)}} (\delta\theta^{(i)}\theta^{(i)}\dot{\eta}^{(i)} - \boldsymbol{Q}^{(i)} \cdot \nabla(\delta\theta^{(i)}))\,\mathrm{d}V - \sum_i \int\limits_{\mathcal{B}_0^{(i)}} \delta\theta^{(i)} R^{(i)} \,\mathrm{d}V \\ - \sum_i \int\limits_{\partial\mathcal{B}_0^{(i),Q}} \delta\theta^{(i)} \bar{Q}_\mathrm{N}^{(i)} \,\mathrm{d}A + \int\limits_{\partial\mathcal{B}_0^{(1),c}} (\delta\theta^{(1)} Q_\mathrm{c}^{(1)} + \delta\theta^{(2)} Q_\mathrm{c}^{(2)})\,\mathrm{d}A = 0\end{aligned} \tag{5.34}$$

3. Interface conditions
   - Normal contact

$$g_\mathrm{N} \geq 0, \qquad t_\mathrm{N} \leq 0, \qquad t_\mathrm{N} g_\mathrm{N} = 0 \tag{5.35}$$

   - Tangential contact

$$\hat{\phi}_\mathrm{c} = \|\boldsymbol{t}_\mathrm{T}\| - \mu_\mathrm{c}|t_\mathrm{N}| \leq 0, \qquad \dot{\zeta} \geq 0, \qquad \hat{\phi}_\mathrm{c}\dot{\zeta} = 0, \qquad \dot{\boldsymbol{g}}_\mathrm{T} = \dot{\zeta}\frac{\boldsymbol{t}_\mathrm{T}}{\|\boldsymbol{t}_\mathrm{T}\|} \tag{5.36}$$

   - Thermal contact

$$Q_\mathrm{c}^{(1)} = k_\mathrm{c}|t_\mathrm{N}|g_\theta - \gamma_\mathrm{c}^{(1)} D_\mathrm{c}^\mathrm{fric}, \qquad Q_\mathrm{c}^{(2)} = -k_\mathrm{c}|t_\mathrm{N}|g_\theta - \gamma_\mathrm{c}^{(2)} D_\mathrm{c}^\mathrm{fric} \tag{5.37}$$

The global energy balance equation given by (3.9) can be augmented by (5.32) as

$$\sum_i \frac{\mathrm{d}}{\mathrm{d}t} \int_{\mathcal{B}_0^{(i)}} \left(\frac{1}{2}\rho_0^{(i)} \dot{\boldsymbol{\varphi}}^{(i)} \cdot \dot{\boldsymbol{\varphi}}^{(i)} + e^{(i)}\right) \mathrm{d}V + \int_{\partial\mathcal{B}_0^{(1),c}} \dot{e}_c \,\mathrm{d}A - \sum_i \int_{\partial\mathcal{B}_0^{(i)}} (\boldsymbol{T}^{(i)} \cdot \dot{\boldsymbol{\varphi}}^{(i)} + Q_{\mathrm{N}}^{(i)})\,\mathrm{d}A$$
$$- \sum_i \int_{\mathcal{B}_0^{(i)}} (\boldsymbol{B}^{(i)} \cdot \dot{\boldsymbol{\varphi}}^{(i)} + R^{(i)})\mathrm{d}V - \int_{\partial\mathcal{B}_0^{(1),c}} (t_{\mathrm{N}}\dot{g}_{\mathrm{N}} + \boldsymbol{t}_{\mathrm{T}} \cdot \dot{\boldsymbol{g}}_{\mathrm{T}} + Q_{\mathrm{c}}^{(1)} + Q_{\mathrm{c}}^{(2)})\,\mathrm{d}A = 0. \tag{5.38}$$

Accordingly, for the variational statements given by (3.50) follows

$$\begin{aligned} \sum_i G_{\varphi}^{(i)} + G_{\varphi}^{\mathrm{c}} &= 0, \\ \sum_i G_{\theta}^{(i)} + G_{\theta}^{\mathrm{c}} &= 0, \end{aligned} \tag{5.39}$$

where the mechanical and thermal contact contributions are given as

$$G_{\varphi}^{\mathrm{c}} = -\sum_i \int_{\partial\mathcal{B}_0^{(i),c}} \delta\boldsymbol{\varphi}^{(i)} \cdot \boldsymbol{t}^{(i)}\mathrm{d}A = \int_{\partial\mathcal{B}_0^{(1),c}} (t_{\mathrm{N}}\delta g_{\mathrm{N}} + \boldsymbol{t}_{\mathrm{T}} \cdot \delta\boldsymbol{g}_{\mathrm{T}})\mathrm{d}A \tag{5.40}$$

and

$$G_{\theta}^{\mathrm{c}} = \sum_i \int_{\partial\mathcal{B}_0^{(i),c}} \delta\theta^{(i)} Q_{\mathrm{c}}^{(i)}\mathrm{d}A = \int_{\partial\mathcal{B}_0^{(1),c}} (\delta\theta^{(1)} Q_{\mathrm{c}}^{(1)} + \delta\theta^{(2)} Q_{\mathrm{c}}^{(2)})\mathrm{d}A. \tag{5.41}$$

Therein, δg_{N} and $\delta \boldsymbol{g}_{\mathrm{T}}$ are the virtual variations of the gap in normal and tangential direction, respectively. The resulting variational formulation and the constitutive contact laws are summarized in Table 5.1.

5.4 Discrete setting

In this section we consider the spatial and temporal discretization of the above defined contact relations. In particular, the spatial discretization of the bodies in contact is carried out using the IGA concept as introduced in chapter 2. For the discrete contact interface we apply a variationally consistent mortar formulation, where linear Lagrangian shape functions are used for the discrete traction field. Moreover, we introduce newly defined triple mortar integrals to take care for a proper energy transfer between the mechanical and the thermal field. Again, for the temporal discretization a mid-point evaluation is applied with respect to the primal variables and the frictional traction field, whereas the mortar integrals are evaluated at time step t_n and the Lagrange multipliers are constant throughout the time step, cf. Dittmann et al. [27].

5.4.1 Mortar method

To construct a mortar based approach for the thermomechanical contact interface, the variational statements in (5.40) and (5.41) have to be discretized as well. Therefore, the space of admissible test functions for the discrete traction field is introduced as

$$\mathcal{M}^{\mathrm{h}} = \{\delta \boldsymbol{t}^{(1),\mathrm{h}} \in \mathcal{L}^2(\partial\mathcal{B}_0^{(1),\mathrm{c}} \cap \partial\mathcal{B}_0^{(2),\mathrm{c}})\}. \tag{5.42}$$

To deal with non-conforming discretizations and hierarchical refined quadratic NURBS, we evaluate the geometrical map $\mathfrak{F}(\boldsymbol{\xi})$ at the boundary of the parameter space, corresponding to the contact boundary in physical space, to obtain a set of physical nodes $\boldsymbol{x}_A^{(1)}$ on $\partial\mathcal{B}^{(1),\mathrm{c}}$ with $A \in \mathcal{I}_{\mathrm{c}} = [1, \ldots, \mathfrak{n}_{\mathrm{c}}]$, where $\mathfrak{n}_{\mathrm{c}}$ corresponds to the number of nodes. As shown in Hesch & Betsch [41], a linear interpolation of the contact traction and their variation

$$\boldsymbol{t}^{(1),\mathrm{h}} = \sum_{A\in\mathcal{I}_{\mathrm{c}}} N^A \boldsymbol{t}_A \qquad \text{and} \qquad \delta\boldsymbol{t}^{(1),\mathrm{h}} = \sum_{A\in\mathcal{I}_{\mathrm{c}}} N^A \delta\boldsymbol{t}_A, \tag{5.43}$$

where $N^A : \partial\mathcal{B}_0^{(1),c} \to \mathbb{R}$ denote $(d-1)$ dimensional linear Lagrangian shape functions associated with nodes A in physical space, is sufficient. Again, decomposing the tractions $\boldsymbol{t}_A$ into normal $\lambda_{\mathrm{N},A}$ (Lagrange multiplier) and tangential $\boldsymbol{t}_{\mathrm{T},A}$ components as shown in (5.9) and subsequently inserting them into (5.40) yields

$$G_\varphi^{\mathrm{c,h}} = \lambda_{\mathrm{N},A}\boldsymbol{n} \cdot \left[n^{AB}\delta\boldsymbol{q}_B^{(1)} - n^{AC}\delta\boldsymbol{q}_C^{(2)}\right] + \boldsymbol{t}_{\mathrm{T},A} \cdot (\boldsymbol{I} - \boldsymbol{n}\otimes\boldsymbol{n})\left[n^{AB}\delta\boldsymbol{q}_B^{(1)} - n^{AC}\delta\boldsymbol{q}_C^{(2)}\right], \tag{5.44}$$

whereas the discrete setting of the contact contribution in (5.41) reads

$$\begin{aligned} G_\theta^{\mathrm{c,h}} = {} & \delta\Theta_A^{(1)} \Big\{ k_{\mathrm{c}}|\lambda_{\mathrm{N},B}| \left[m^{ABC}\Theta_C^{(1)} - m^{ABD}\Theta_D^{(2)}\right] \\ & \quad - \gamma_{\mathrm{c}}^{(1)}\boldsymbol{t}_{\mathrm{T},B} \cdot (\boldsymbol{I} - \boldsymbol{n}\otimes\boldsymbol{n}) \left[m^{ABC}\dot{\boldsymbol{q}}_C^{(1)} - m^{ABD}\dot{\boldsymbol{q}}_D^{(2)}\right]\Big\} \\ & + \delta\Theta_A^{(2)} \Big\{ -k_{\mathrm{c}}|\lambda_{\mathrm{N},B}| \left[\bar{m}^{ABC}\Theta_C^{(1)} - \bar{m}^{ABD}\Theta_D^{(2)}\right] \\ & \quad - \gamma_{\mathrm{c}}^{(2)}\boldsymbol{t}_{\mathrm{T},B} \cdot (\boldsymbol{I} - \boldsymbol{n}\otimes\boldsymbol{n}) \left[\bar{m}^{ABC}\dot{\boldsymbol{q}}_C^{(1)} - \bar{m}^{ABD}\dot{\boldsymbol{q}}_D^{(2)}\right]\Big\}. \end{aligned} \tag{5.45}$$

Here, the mortar integrals are defined as follows

$$\begin{aligned} n^{AB} &= \int_{\partial\mathcal{B}_0^{(1),c}} N^A(\boldsymbol{\xi}^{(1)}) R^B(\boldsymbol{\xi}^{(1)}) \,\mathrm{d}A, \\ n^{AC} &= \int_{\partial\mathcal{B}_0^{(1),c}} N^A(\boldsymbol{\xi}^{(1)}) R^C(\boldsymbol{\xi}^{(2)}) \,\mathrm{d}A \end{aligned} \tag{5.46}$$

and the triple mortar integrals are given by

$$\begin{aligned} m^{ABC} &= \int_{\partial\mathcal{B}_0^{(1),c}} R^A(\boldsymbol{\xi}^{(1)}) N^B(\boldsymbol{\xi}^{(1)}) R^C(\boldsymbol{\xi}^{(1)}) \,\mathrm{d}A, \\ m^{ABD} &= \int_{\partial\mathcal{B}_0^{(1),c}} R^A(\boldsymbol{\xi}^{(1)}) N^B(\boldsymbol{\xi}^{(1)}) R^D(\boldsymbol{\xi}^{(2)}) \,\mathrm{d}A. \end{aligned} \tag{5.47}$$

The evaluation of the triple mortar integrals $\bar{m}$ follows analogously using $R^A(\boldsymbol{\xi}^{(2)})$ as first shape function in (5.47). The contact conditions required for the evaluation of the corresponding contact forces and the heat transfer of the semi-discrete setting are summarized in Table 5.2.

Table 5.2: Semi-discrete formulation of the thermomechanical contact problem

1. Mechanical field

$$\begin{aligned}&\sum_i \delta \boldsymbol{q}_A^{(i)} \cdot \left[M^{(i),AB} \dot{\boldsymbol{v}}_B^{(i)} + \int_{\mathcal{B}_0^{(i)}} \boldsymbol{P}^{(i),\mathrm{h}} \nabla(R^A)\,\mathrm{d}V - \boldsymbol{F}^{(i),\mathrm{ext},A} \right] \\ &+ \lambda_{\mathrm{N},A} \boldsymbol{n} \cdot \left[n^{AB} \delta \boldsymbol{q}_B^{(1)} - n^{AC} \delta \boldsymbol{q}_C^{(2)} \right] + \boldsymbol{t}_{\mathrm{T},A} \cdot (\boldsymbol{I} - \boldsymbol{n} \otimes \boldsymbol{n}) \left[n^{AB} \delta \boldsymbol{q}_B^{(1)} - n^{AC} \delta \boldsymbol{q}_C^{(2)} \right] = 0\end{aligned} \tag{5.48}$$

2. Thermal field

$$\begin{aligned}&\sum_i \delta \Theta_A^{(i)} \left[R_A \Gamma^{(i),\mathrm{h}}\,\mathrm{d}V - \int_{\mathcal{B}_0^{(i)}} \boldsymbol{Q}^{(i),\mathrm{h}} \cdot \nabla(R^A)\,\mathrm{d}V - \bar{Q}^{(i),A} \right] \\ &+ \delta \Theta_A^{(1)} \Big\{ k_{\mathrm{c}} |\lambda_{\mathrm{N},B}| \left[m^{ABC} \Theta_C^{(1)} - m^{ABD} \Theta_D^{(2)} \right] \\ &\qquad - \gamma_{\mathrm{c}}^{(1)} \boldsymbol{t}_{\mathrm{T},B} \cdot (\boldsymbol{I} - \boldsymbol{n} \otimes \boldsymbol{n}) \left[m^{ABC} \dot{\boldsymbol{q}}_C^{(1)} - m^{ABD} \dot{\boldsymbol{q}}_D^{(2)} \right] \Big\} \\ &+ \delta \Theta_A^{(2)} \Big\{ -k_{\mathrm{c}} |\lambda_{\mathrm{N},B}| \left[\bar{m}^{ABC} \Theta_C^{(1)} - \bar{m}^{ABD} \Theta_D^{(2)} \right] \\ &\qquad - \gamma_{\mathrm{c}}^{(2)} \boldsymbol{t}_{\mathrm{T},B} \cdot (\boldsymbol{I} - \boldsymbol{n} \otimes \boldsymbol{n}) \left[\bar{m}^{ABC} \dot{\boldsymbol{q}}_C^{(1)} - \bar{m}^{ABD} \dot{\boldsymbol{q}}_D^{(2)} \right] \Big\} = 0\end{aligned} \tag{5.49}$$

3. Semi-discrete contact conditions
   - Normal contact

$$g_{\mathrm{N}}^A \geq 0, \qquad \lambda_{\mathrm{N},A} \leq 0, \qquad \lambda_{\mathrm{N},A} g_{\mathrm{N}}^A = 0, \qquad g_{\mathrm{N}}^A = \boldsymbol{n} \cdot \left[n^{AB} \boldsymbol{q}_B^{(1)} - n^{AC} \boldsymbol{q}_C^{(2)} \right] \tag{5.50}$$

   - Tangential contact

$$\hat{\phi}_{\mathrm{c},A} = \|\boldsymbol{t}_{\mathrm{T},A}\| - \mu_{\mathrm{c}} |\lambda_{\mathrm{N},A}| \leq 0, \qquad \dot{\zeta}^A \geq 0, \quad \hat{\phi}_{\mathrm{c},A} \dot{\zeta}^A = 0, \quad \dot{\boldsymbol{g}}_{\mathrm{T}}^A = \dot{\zeta}^A \frac{\boldsymbol{t}_{\mathrm{T},A}}{\|\boldsymbol{t}_{\mathrm{T},A}\|} \tag{5.51}$$

with

$$\dot{\boldsymbol{g}}_{\mathrm{T}}^A = (\boldsymbol{I} - \boldsymbol{n} \otimes \boldsymbol{n}) \left[n^{AB} \dot{\boldsymbol{q}}_B^{(1)} - n^{AC} \dot{\boldsymbol{q}}_C^{(2)} \right] \tag{5.52}$$

For the numerical evaluation of the mortar integrals, a suitable segmentation process is necessary, subdividing the surface of the parameter space on both sides in triangles. We focus on the evaluation of the standard mortar integrals, since the triple mortar integrals follow analogously, extended by additional shape functions R^A. The main goal of this construction is to provide a common parametrization of both surfaces to apply Gauss quadrature. Therefore, a linear transformation based on bilinear, triangular shape functions M^K is introduced as

$$\boldsymbol{\xi}^{(i),\mathrm{h}}(\boldsymbol{\eta}) = \sum_{K=1}^{3} M^K(\boldsymbol{\eta})\boldsymbol{\xi}_K^{(i)}, \tag{5.53}$$

such that we obtain the segment contributions of the mortar integrals

$$\begin{aligned} n^{\kappa\beta} &= \int_{\triangle} N^{\kappa}(\boldsymbol{\xi}^{(1),\mathrm{h}}(\boldsymbol{\eta}))R^{\beta}(\boldsymbol{\xi}^{(1),\mathrm{h}}(\boldsymbol{\eta}))J_{\mathrm{seg}}\,\mathrm{d}A_{\triangle}, \\ n^{\kappa\zeta} &= \int_{\triangle} N^{\kappa}(\boldsymbol{\xi}^{(1),\mathrm{h}}(\boldsymbol{\eta}))R^{\zeta}(\boldsymbol{\xi}^{(2),\mathrm{h}}(\boldsymbol{\eta}))J_{\mathrm{seg}}\,\mathrm{d}A_{\triangle}, \end{aligned} \tag{5.54}$$

which have to be assembled into the global system, see Hesch & Betsch [37, 40] and the references therein. Note that the triangle symbols in equation (5.54) represent the triangle reference element. Finally, the Jacobian of each segment is required

$$J_{\mathrm{seg}} = \|\boldsymbol{A}_1(\boldsymbol{\xi}^{(1),\mathrm{h}}(\boldsymbol{\eta})) \times \boldsymbol{A}_2(\boldsymbol{\xi}^{(1),\mathrm{h}}(\boldsymbol{\eta}))\| \det(\mathrm{D}\boldsymbol{\xi}(\boldsymbol{\eta})), \tag{5.55}$$

where $\boldsymbol{A}_\alpha(\boldsymbol{\xi}) = \frac{\mathrm{d}R^A(\boldsymbol{\xi})}{\mathrm{d}\xi_\alpha}\boldsymbol{X}_A$ with $\alpha = \{1, 2\}$ denote the tangential vectors in the reference configuration.

To construct the segments by calculating the nodes $\boldsymbol{\xi}_K^{(i)}$ within the parameter space, a stable, but tedious method has been proposed in Hesch & Betsch [41]. Here, we propose a major simplification of the segmentation algorithm for isogeometric discretized bodies, which allows the usage of standard segmentation libraries, cf. Dittmann et al. [27]. Therefore, we reuse the already introduced set of nodes contained in $\mathcal{I}_{\mathrm{c}}$ and apply a linear Lagrangian discretization in analogy to the linear interpolation of the contact traction field in (5.43). Based on this artificially discretized, virtual surface and a corresponding virtual discretization of the contact boundary of $\mathcal{B}^{(2)}$, we execute the standard segmentation procedure, see Figure 5.3 for a virtual segmentation surface on an arbitrary curved B-spline surface.

Remark 6: We use this simplification only for the segmentation process, i.e. for the calculation of the local coordinates $\boldsymbol{\xi}_K^{(i)}$ and not for the evaluation of the mortar integrals. The speed up is tremendous, furthermore, the definition of the segment corners is not as strict as the definition of the closest point projection used within a collocation method, see Temizer [92]. More importantly, the proposed method yields a set of valid, non-overlapping segments, we only shift small area contributions from one segment to the next.

Remark 7: Due to the presence of quadratic B-splines and NURBS, the chosen Gauss quadrature has to be adjusted carefully. A standard four point integration is not sufficient, especially for the evaluation of the triple mortar integrals.

Remark 8: The variation of the normal gap δg_{N} for an initially curved surface may include additionally the variation of the normal vector, which is dropped for convenience in equation (5.44) and the following. This is sometimes done in computational contact mechanics based on the mortar method (see e.g. Laursen et al. [57]). In contrast to collocation methods, where the variation of the normal vector drops out (see e.g. Wriggers [99]), the normal components for mortar methods are assembled from all involved segments, such that the corresponding variation does not vanish in all situations. Leaving the variation out of the balance equation may have considerable impact to the angular momentum at the contact interface. This has already been observed in Puso and Laursen [76, footnote p. 606] and Puso and Laursen [77, footnote 1 and 2 p. 4896].

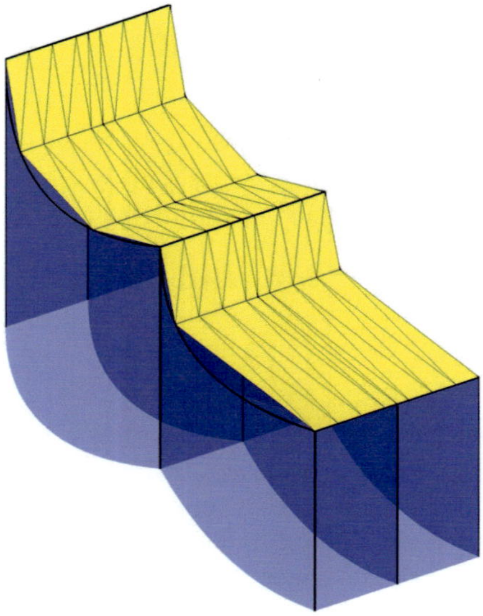

Figure 5.3: Virtual segmentation surface (yellow) on an arbitrary curved B-spline geometry (blue).

Remark 9: The discrete Lagrange multiplier space for contact in normal direction has to be constructed with care to avoid possible singularities. To clarify this issue, Figure 5.4 shows a representative example for quadratic B-splines. The unrefined situation is displayed on the left hand side, consisting of 49 shape functions on the particular surface with 36 physical nodes. On the right hand side, a single B-spline has been refined (all remaining shape functions are transparent in this figure). Here, we obtain 64 shape functions with associated degrees of freedom, whereas 69 physical nodes on the surface are given. Consideration of a low order Lagrange multiplier space, e.g. a Dirac distribution of the multipliers at the physical nodes would clearly overconstrain the system. This issue arises for constraints like Dirichlet boundary conditions or for incompressibility constraints. We did not observe this for the contact constraints, although this cannot be ruled out by design. A simple remedy can be considered: Since this issue arises only in the transfer area between coarse and fully refined mesh, the Lagrange multipliers can be placed on the coarse level elements, reducing the total number of constraints.

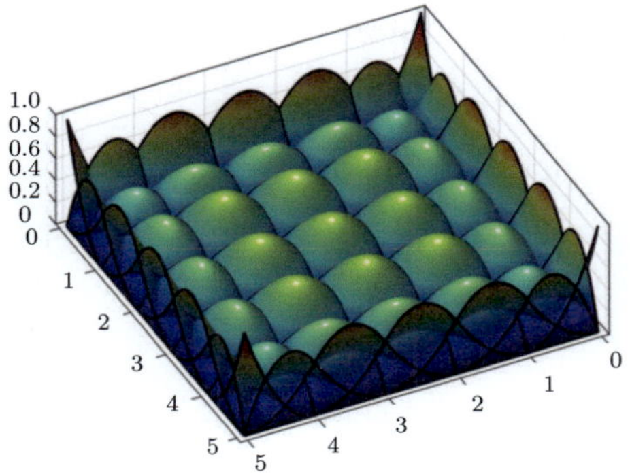

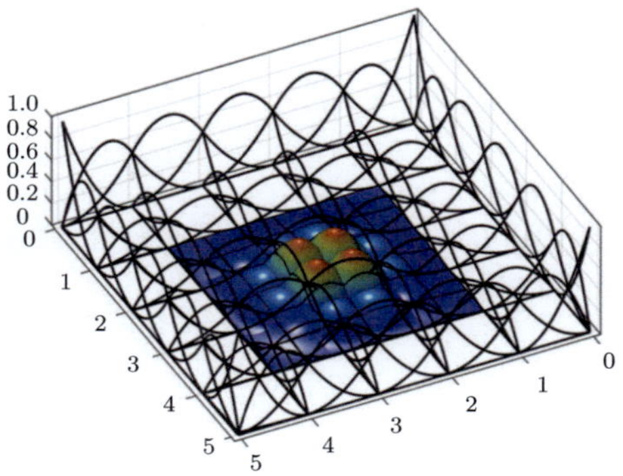

Figure 5.4: Unrefined B-spline basis (left) and refined shape functions (right).

5.4.2 Temporal discretization

Following the approach outlined in Hesch & Betsch [40], we evaluate the mortar integrals at time t_n, whereas the Lagrange multipliers $\lambda_{\mathrm{N},A,n+1}$ remain constant throughout the time step. The Karush-Kuhn-Tucker conditions (5.50) are implemented via an active set strategy and the weak formulation reads

$$G^{\mathrm{c,h}}_{\varphi,\mathrm{N}} = \lambda_{\mathrm{N},A,n+1}\boldsymbol{n}_{n+1/2} \cdot \left[n^{AB}_n \delta\boldsymbol{q}^{(1)}_B - n^{AC}_n \delta\boldsymbol{q}^{(2)}_C\right]. \tag{5.56}$$

Note that we evaluate the corresponding constraints at time t_{n+1}.

The tangential contribution to the weak formulation follows analogously

$$G^{\mathrm{c,h}}_{\varphi,\mathrm{T}} = \boldsymbol{t}_{\mathrm{T},A,n+1/2} \cdot (\boldsymbol{I} - \boldsymbol{n}_{n+1/2} \otimes \boldsymbol{n}_{n+1/2}) \left[n^{AB}_n \delta\boldsymbol{q}^{(1)}_B - n^{AC}_n \delta\boldsymbol{q}^{(2)}_C\right]. \tag{5.57}$$

In particular, we employ a trial state-return map strategy to determine the Coulomb frictional traction. The trial state is given by

$$\boldsymbol{t}^{\mathrm{trial}}_{\mathrm{T},A,n+1} = \boldsymbol{t}_{\mathrm{T},A,n} + \epsilon_{\mathrm{T}}(\boldsymbol{I} - \boldsymbol{n}_{n+1/2} \otimes \boldsymbol{n}_{n+1/2}) \left[n^{AB}_n \Delta\boldsymbol{q}^{(1)}_B - n^{AC}_n \Delta\boldsymbol{q}^{(2)}_C\right], \tag{5.58}$$

where $\Delta\boldsymbol{q}^{(1)}_B = \boldsymbol{q}^{(1)}_{B,n+1} - \boldsymbol{q}^{(1)}_{B,n}$ and $\Delta\boldsymbol{q}^{(2)}_C = \boldsymbol{q}^{(2)}_{C,n+1} - \boldsymbol{q}^{(2)}_{C,n}$. To prevent large errors in angular momentum, a modification of the form

$$\begin{aligned}\boldsymbol{t}^{\mathrm{trial}}_{\mathrm{T},A,n+1} =& \boldsymbol{t}_{\mathrm{T},A,n} \\ &+ \epsilon_{\mathrm{T}}(\boldsymbol{I} - \boldsymbol{n}_{n+1/2} \otimes \boldsymbol{n}_{n+1/2}) \left[(n^{AB}_{n+1} - n^{AB}_n)\boldsymbol{q}^{(1)}_{B,n+1/2} - (n^{AC}_{n+1} - n^{AC}_n)\boldsymbol{q}^{(2)}_{C,n+1/2}\right]\end{aligned} \tag{5.59}$$

is often used. For both approaches, the slip function can be obtained for each physical node A separately

$$\hat{\phi}_{\mathrm{c},A,n+1} = \|\boldsymbol{t}^{\mathrm{trial}}_{\mathrm{T},A,n+1}\| - \mu_{\mathrm{c}}|\lambda_{\mathrm{N},A,n+1}| \tag{5.60}$$

and we obtain the frictional tractions

$$\boldsymbol{t}_{\mathrm{T},A,n+1} = \begin{cases} \boldsymbol{t}^{\text{trial}}_{\mathrm{T},A,n+1}, & \text{if } \hat{\phi}_{\mathrm{c},A,n+1} \leq 0 \\ \mu_{\mathrm{c}} |\lambda_{\mathrm{N},A,n+1}| \frac{\boldsymbol{t}^{\text{trial}}_{\mathrm{T},A,n+1}}{\|\boldsymbol{t}^{\text{trial}}_{\mathrm{T},A,n+1}\|}, & \text{if } \hat{\phi}_{\mathrm{c},A,n+1} > 0 \end{cases}. \tag{5.61}$$

Once the frictional tractions at time t_{n+1} are calculated, the corresponding midpoint approximation reads

$$\boldsymbol{t}_{\mathrm{T},A,n+1/2} = \frac{1}{2}(\boldsymbol{t}_{\mathrm{T},A,n+1} + \boldsymbol{t}_{\mathrm{T},A,n}), \tag{5.62}$$

where we follow the arguments outlined in Armero & Petöcz [5].

Eventually, the thermal contribution to the weak formulation is given by

$$\begin{aligned} G^{\mathrm{c,h}}_{\theta} = \delta\Theta^{(1)}_A \Bigg\{ & k_{\mathrm{c}} |\lambda_{\mathrm{N},B,n+1}| \left[m^{ABC}_n \Theta^{(1)}_{C,n+1/2} - m^{ABD}_n \Theta^{(2)}_{D,n+1/2} \right] \\ & - \gamma^{(1)}_{\mathrm{c}} \boldsymbol{t}_{\mathrm{T},B,n+1} \cdot (\boldsymbol{I} - \boldsymbol{n}_{n+1/2} \otimes \boldsymbol{n}_{n+1/2}) \left[m^{ABC}_n \frac{\Delta \boldsymbol{q}^{(1)}_C}{\Delta t} - m^{ABD}_n \frac{\Delta \boldsymbol{q}^{(2)}_D}{\Delta t} \right] \Bigg\} \\ + \delta\Theta^{(2)}_A \Bigg\{ & -k_{\mathrm{c}} |\lambda_{\mathrm{N},B,n+1}| \left[\bar{m}^{ABC}_n \Theta^{(1)}_{C,n+1/2} - \bar{m}^{ABD}_n \Theta^{(2)}_{D,n+1/2} \right] \\ & - \gamma^{(2)}_{\mathrm{c}} \boldsymbol{t}_{\mathrm{T},B,n+1} \cdot (\boldsymbol{I} - \boldsymbol{n}_{n+1/2} \otimes \boldsymbol{n}_{n+1/2}) \left[\bar{m}^{ABC}_n \frac{\Delta \boldsymbol{q}^{(1)}_C}{\Delta t} - \bar{m}^{ABD}_n \frac{\Delta \boldsymbol{q}^{(2)}_D}{\Delta t} \right] \Bigg\} \end{aligned} \tag{5.63}$$

assuming that (5.58) is used. The modified form in (5.59) can be applied using $\frac{\Delta m^{ABC}}{\Delta t}$ and $\frac{\Delta m^{ABD}}{\Delta t}$ instead of $\frac{\Delta \boldsymbol{q}^{(1)}_C}{\Delta t}$ and $\frac{\Delta \boldsymbol{q}^{(2)}_D}{\Delta t}$ within the last expression.

Remark 10: The above thermomechanical contact formulation can be adapted to fracture mechanical contact problems by neglecting the thermal contact contribution, see Hesch et al. [42]. Then, the contact surface $\partial\mathcal{B}^{(i),\mathrm{c}}_0$ does not interfere with the phase-field boundary, which is in contrast to the thermal boundary of the thermomechanical problem, where an energy transfer across the contact interface is established.

5.5 Numerical examples

To demonstrate the accuracy and performance of the proposed methodology for thermomechanical and fracture mechanical contact problems we investigate several quasi-static as well as transient numerical examples.

5.5.1 Thermomechanical contact problems

We start this section with a modified patch test to outline the capabilities of the thermomechanical mortar method for IGA. In a second example, we focus our investigation on the interaction between the mechanical and the thermal field using a quasi-static ironing example. Finally, a transient impact simulation demonstrates the accuracy of the frictional mortar contact algorithm in terms of the first and second law of thermodynamics.

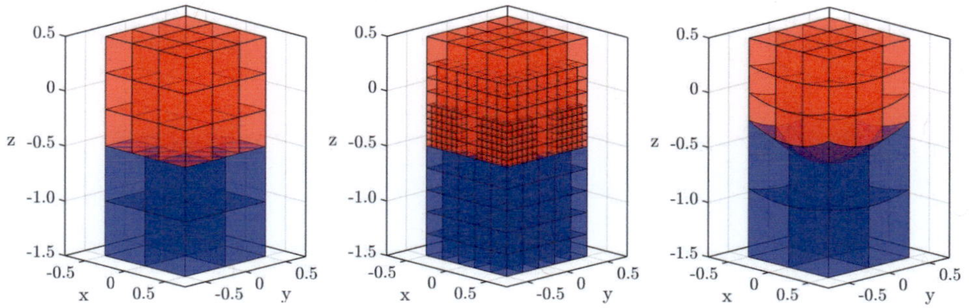

Figure 5.5: Patch test: Reference configuration with flat (left), flat refined (middle) and curved (right) interface.

The material behavior within the three examples is assumed to be governed by the Mooney-Rivlin material law given by (3.29), where the corresponding material parameters are specified in each example using SI-units. Again, the chosen values for the material parameters do not correspond to any physical material, i.e. the examples presented in this section are of purely academic nature.

Patch test

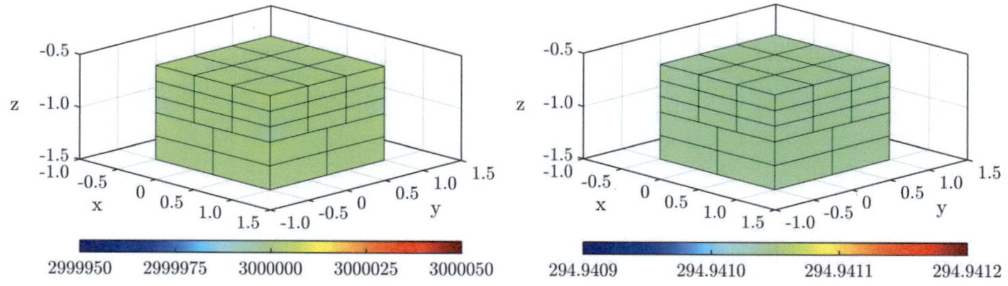

Figure 5.6: Patch test: Von Mises stress distribution (left) and temperature distribution (right) for two non-conform meshed blocks.

First, we investigate a modified patch test with flat as well as curved interfaces. As shown in Figure 5.5, two independently meshed blocks are tied together via the proposed mortar method, assuming that both bodies remain in contact without tangential sliding and without thermal resistance at the interface. A Neumann boundary is applied to the top surface, whereas the bottom surface is clamped, such that the body can expand in horizontal direction. For the discretization of both blocks we apply quadratic B-spline based shape functions with different numbers of elements, see Figure 5.5.

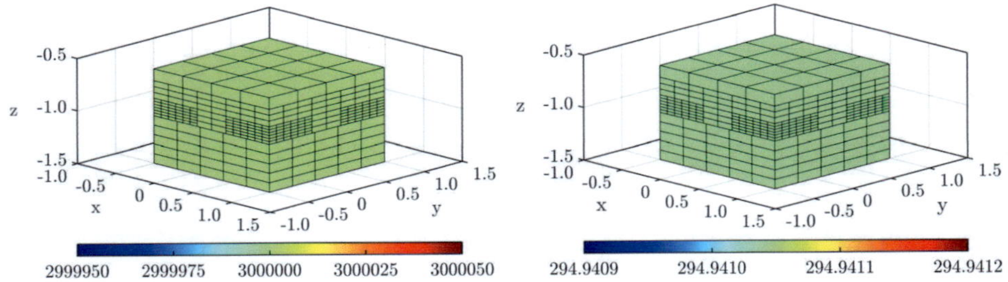

Figure 5.7: Patch test: Von Mises stress distribution (left) and temperature distribution (right) for a hierarchical refined contact interface.

The material setting is identical to the setting used for the patch test presented in chapter 3, i.e. the values for the mechanical material parameters are $\alpha = 15/13 \times 10^5$, $\beta = 10/13 \times 10^5$ and $\kappa = 25/3 \times 10^5$, whereas the setting for the thermal field reads $c = 1830$, $\gamma = 0.22333$ and $K = K_{\text{ref}} = 0.55$. As before, the mass density is set to zero and the initial temperature is 293.15 for both blocks.

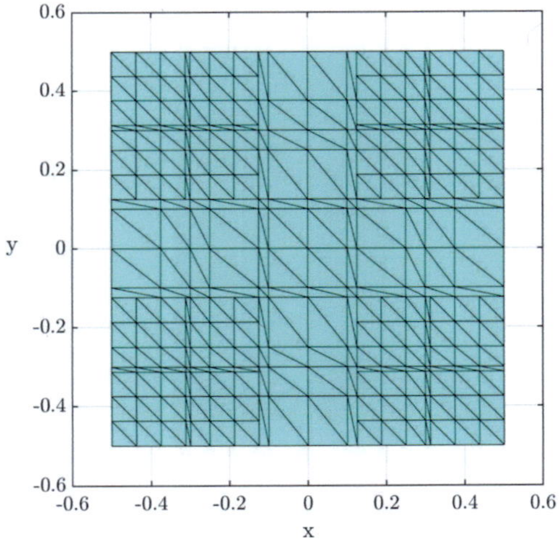

Figure 5.8: Patch test: Segmentation of the refined interface.

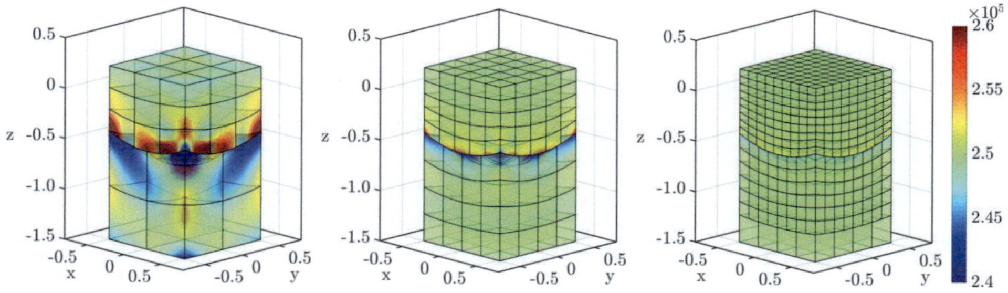

Figure 5.9: Patch test: Von Mises stress distribution for different discretizations of the curved contact interface.

Figure 5.6 (left) shows the von Mises stress distribution of the solution for a uniform pressure field of $p = 3 \times 10^6$, applied to the upper block. In addition, we obtain a nearly perfect and uniform temperature distribution for the proposed linear approximation of the Lagrange multiplier field, see Figure 5.6 (right).

Next, we demonstrate the applicability of the mortar method in conjunction with hierarchical refinements on the interface. Therefore, a two level hierarchical refinement is applied to the lower surface of the upper block, see Figure 5.5 (middle). Again, nearly perfect and uniform results are obtained for the von Mises stress distribution and temperature distribution as depicted in Figure 5.7. The corresponding segmentation of the interface is plotted on the refined surface of the upper block in Figure 5.8.

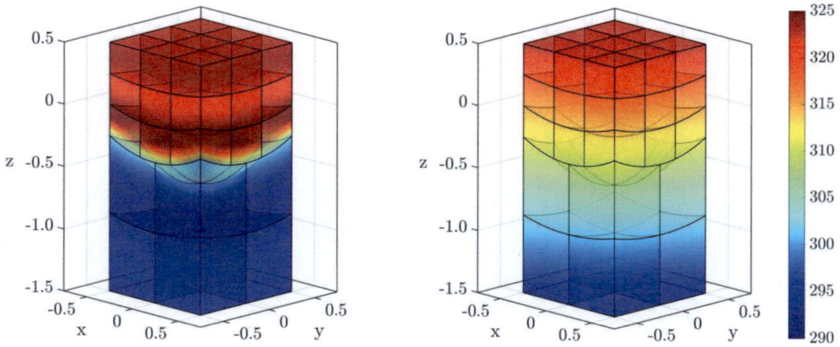

Figure 5.10: Patch test: Thermal distribution after the first step (left) and converged quasi-static solution (right).

For further investigations, a highly curved interface is applied, see Figure 5.5 (right). Note that quadratic B-splines are able to represent the geometry exactly, independent of the chosen number of elements, such that both interfaces are geometrically conform and errors we obtain are discretization errors of the mortar interface. Again, a uniform pressure field of $p = 2.5 \times 10^5$ has been applied to the top surface. The error in the stress field decreases and converges to the correct, uniform solution with higher element numbers at the interface, see Figure 5.9. As already shown in Hesch and Betsch [37], the results of the mortar method for the given curved interface are in general superior to results of collocation type methods.

Since we assume that the error in the stress field is of geometrical nature, we investigate the scalar valued thermal field again. Therefore, the Neumann boundary condition as well as the thermal resistance within the interface are set to zero. A uniform, stress free configuration with initial temperature of $\theta_0 = 323.15$ is predefined for the upper block, whereas an initial temperature of $\theta_0 = 293.15$ is applied to the lower block. The temperature fields at the top and bottom surfaces are fixed. Figure 5.10 (left) shows the solution after a single time step with $\Delta t = 0.05$. Due to the given heat capacity and the thermal conductivity, heat flows across the interface, until the solution converges to the static solution, as shown in Figure 5.10 (right) after a total simulation time of $t = 10^4$.

Ironing problem

In this example, we focus on the interaction between the frictional and the thermal contributions at the interface, using a setup similar to the ironing problem described in Puso and Laursen [77]. The geometry in its reference configuration is shown in Figure 5.11, where the upper block of size $0.9 \times 0.9 \times 0.9$ consists of 27 quadratic B-spline elements and the lower block of size $10 \times 4 \times 3$ consists of 240 quadratic B-spline elements. In addition, a two level hierarchical refinement is applied to the expected contact area on the lower block.

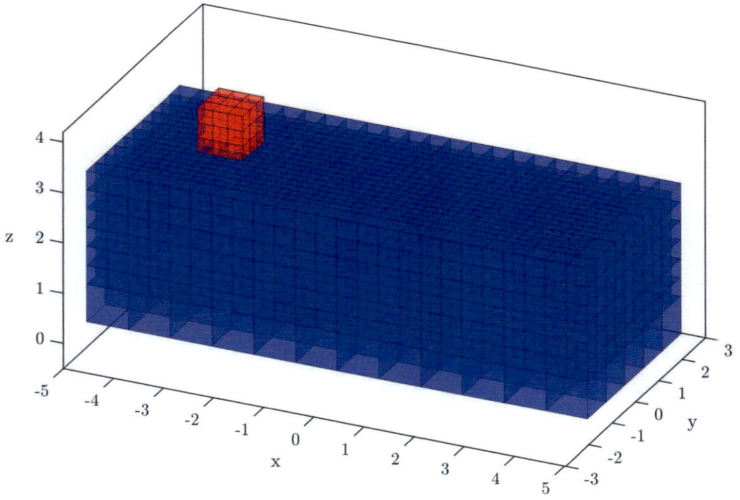

Figure 5.11: Ironing problem: Reference configuration.

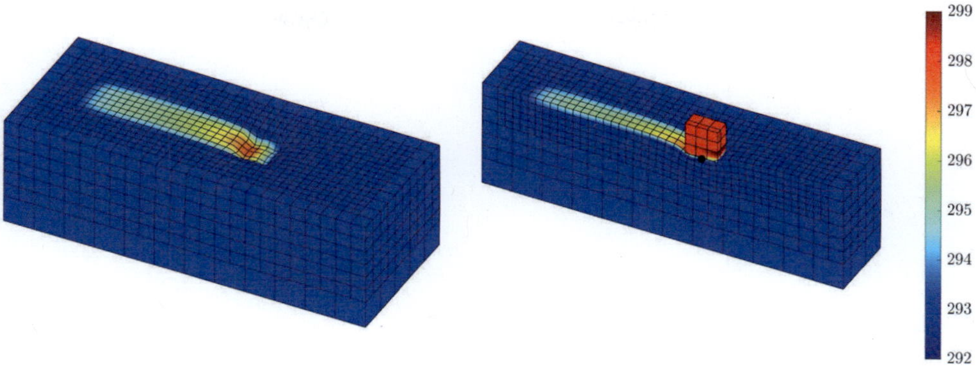

Figure 5.12: Ironing problem: Deformed geometry (left) and sectional view on the symmetry plane (right) at time $t = 0.67$.

The parameters for the Mooney-Rivlin material law are set to $\alpha = 1500/13$, $\beta = 1000/13$ and $\kappa = 2500/3$ for the upper block and $\alpha = 15/13$, $\beta = 10/13$ and $\kappa = 25/3$ for the lower block. The values correspond to a Young's modulus of $E = 10$ and $E = 1000$, respectively and to a Poisson's ratio of $\nu = 0.3$ for both blocks. The setup of the thermal field is given as follows. The initial temperature of the upper block is set to $\theta_0 = 298.15$ and the heat capacity to $c = 1.83 \times 10^4$ in order to avoid a fast cool-down, while the corresponding setting for the lower block is given by $\theta_0 = 293.15$ and $c = 1830$. The thermal conductivity $K = K_{\text{ref}} = 0.55$ and the expansion coefficient $\gamma = 2.2333 \times 10^{-4}$ are set for both blocks. For the contact interface, the effusivities are $\gamma_c^{(1)} = \gamma_c^{(2)} = 0.5$, the heat exchange coefficient is $k_c = 3000$ and the friction coefficient is $\mu_c = 0.2$. Accordingly, frictional dissipation is taken into account.

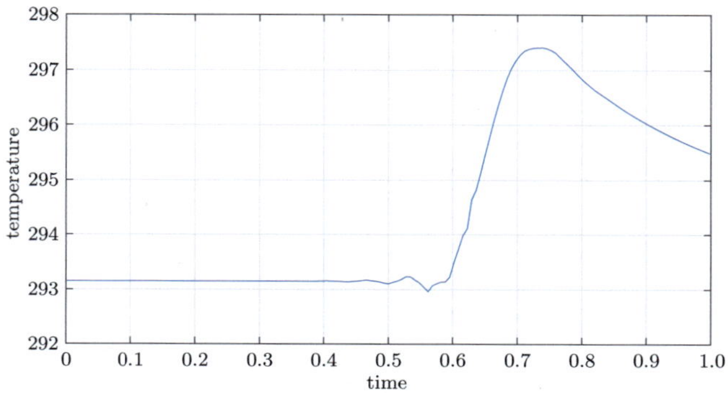

Figure 5.13: Ironing problem: Temperature at the marked position over time.

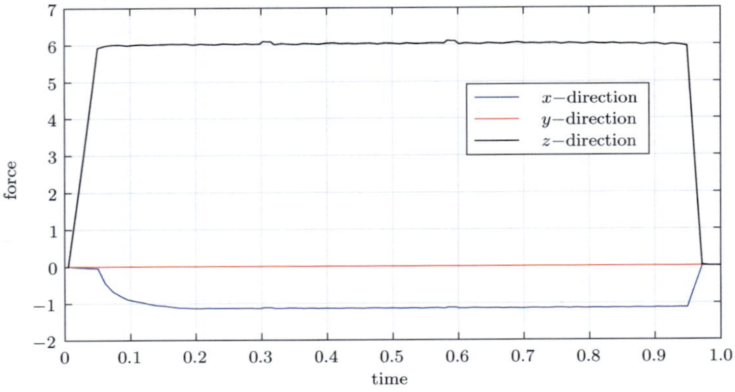

Figure 5.14: Ironing problem: Contact forces acting on the upper block over time for a friction coefficient $\mu_c = 0.2$.

The lower surface of the lower block is fixed, whereas the movement of the upper surface of the upper block is predefined. In particular, the block is first moved downwards (negative z-direction) and pressed into the lower block and then moved sidewards (x-direction), such that it slides over the refined upper surface of the lower block. In Figure 5.12, the temperature distribution at time $t = 0.67$ is shown. On the right hand side, a cut at the symmetry plane of both blocks is shown along with the marking of a specific node. The temperature at this marked position on the physical mesh is plotted over time in Figure 5.13 and the contact forces acting on upper block are plotted over time in Figure 5.14. Due to the frictional contact forces, a small bow wave is running at the front of the upper block with a compression and an expansion part. In connection with the expansion coefficient, this leads to the small temperature wave between $t = 0.45$ and $t = 0.6$. Afterwards, the marked node on the mesh gets in contact with the upper block, which leads to a drastic temperature increase. Note that this example is quite difficult, since we obtain high stresses at the corners of the block.

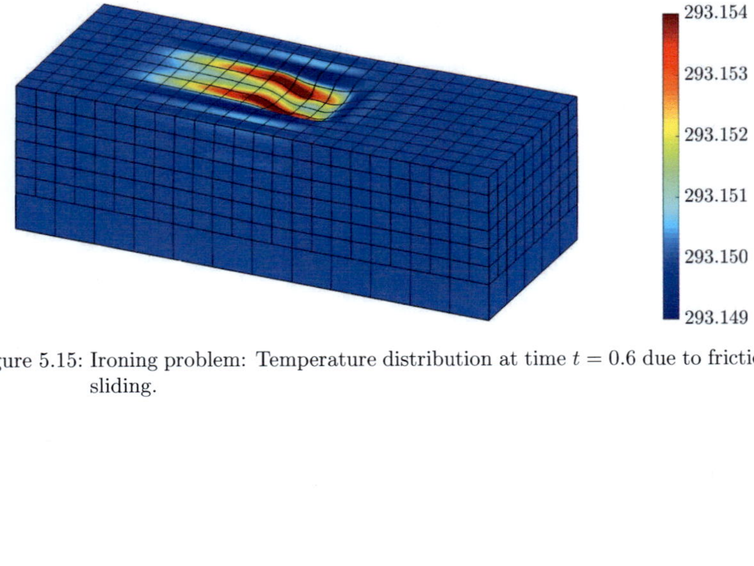

Figure 5.15: Ironing problem: Temperature distribution at time $t = 0.6$ due to frictional sliding.

In the final setup for this problem, we focus on the frictional contributions. Therefore, the thermal expansion coefficient is set to zero and the initial temperature at 293.15 is assumed to be equal for both bodies, i.e. the only heat source is the frictional dissipation, see Figure 5.15 for the resulting temperature distribution.

Impact problem

Next, we deal with a fully coupled transient thermoelastic impact problem of two hollow tori including large deformations, see also Yang & Laursen [101] and Hesch & Betsch [39] for a purely elastodynamic definition of the problem. The reference configuration is shown in Figure 5.16, the inner and outer radius of the torus are 52 and 100, respectively. Moreover, the wall thickness is 4.5 for both tori. For the spatial discretization, 3456 quadratic NURBS elements are used for each torus with 26796 thermal and mechanical degrees of freedom. In its initial configuration the left torus has a homogeneous velocity of $\boldsymbol{v}_0 = [30,\, 0,\, 20]$ and a homogeneous temperature of $\theta_0 = 293.15$, whereas the right torus rests and has a temperature of $\theta_0 = 298.15$.

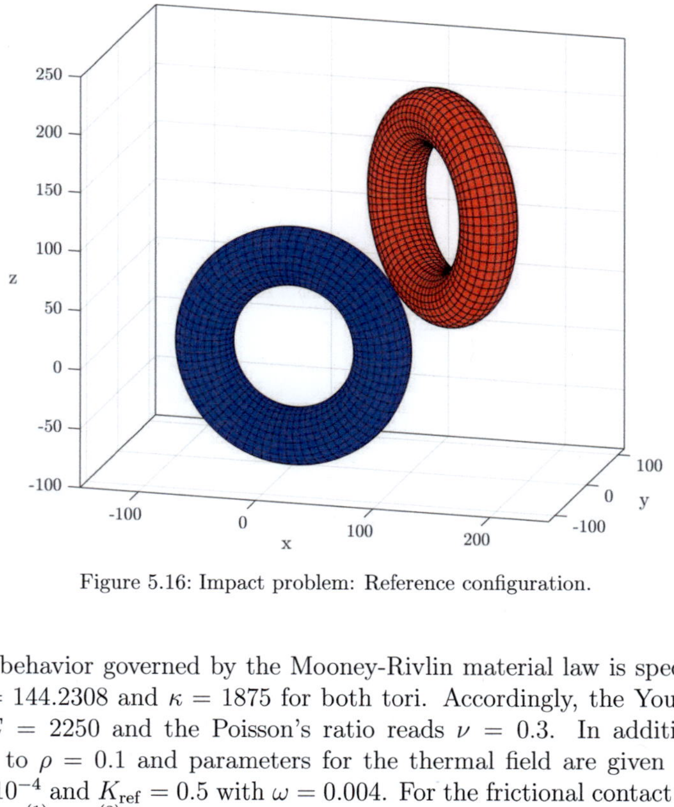

Figure 5.16: Impact problem: Reference configuration.

The material behavior governed by the Mooney-Rivlin material law is specified via $\alpha = 288.4615$, $\beta = 144.2308$ and $\kappa = 1875$ for both tori. Accordingly, the Young's modulus is given by $E = 2250$ and the Poisson's ratio reads $\nu = 0.3$. In addition, the mass density is set to $\rho = 0.1$ and parameters for the thermal field are given by $c = 1830$, $\gamma = 2.2333 \times 10^{-4}$ and $K_{\text{ref}} = 0.5$ with $\omega = 0.004$. For the frictional contact interface, the effusivities are $\gamma_{\text{c}}^{(1)} = \gamma_{\text{c}}^{(2)} = 0.5$, the heat exchange coefficient is $k_{\text{c}} = 100$ and the friction coefficient is $\mu_{\text{c}} = 0.12$.

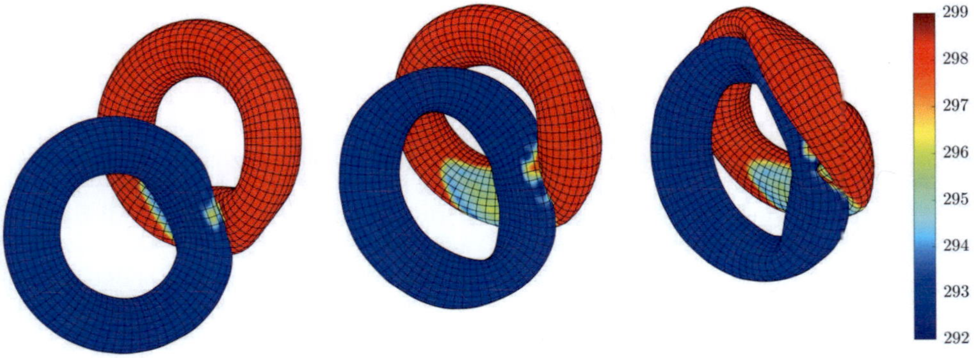

Figure 5.17: Impact problem: Snapshots of the deformed configuration at different times $t = [2,\ 3.5,\ 5]$.

In Figure 5.17, snapshots of the impact for different times are depicted along with the corresponding temperature distribution for both tori. Moreover, the segmentation of the contact interface at time $t = 3.5$ is plotted on the highly curved and deformed surfaces in Figure 5.18. Note that the construction of such a segmentation algorithm is extremely challenging from a technical point of view.

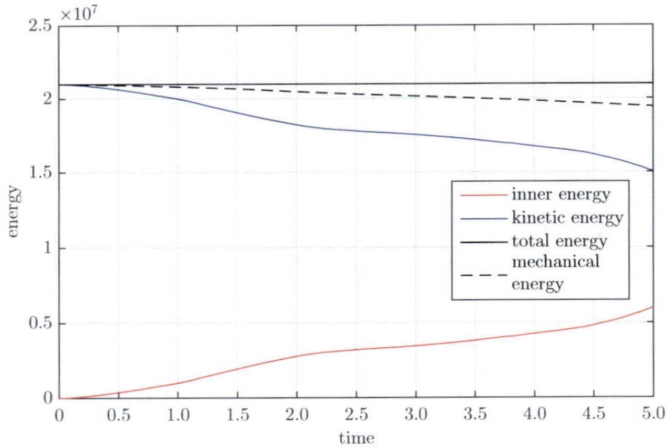

Figure 5.19: Impact problem: Energy plotted over time.

The inner energy, the kinetic energy, the mechanical energy (i.e. the kinetic plus the free Helmholtz energy) as well as the total energy (i.e. the kinetic plus the inner energy) are plotted over time in Figure 5.19 using a constant time step size of $\Delta t = 0.005$. The difference between the mechanical and the total energy represents the dissipated energy, transferred to the thermal field. Furthermore, as requested by the second law of thermodynamics, dissipation is always non-negative.

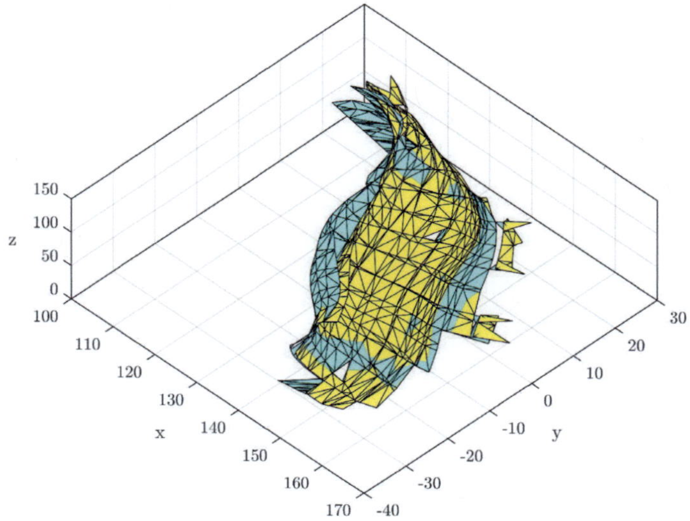

Figure 5.18: Impact problem: Segmentation of the contact interface plotted on both surfaces.

5.5.2 Fracture mechanical contact problems

The purpose of this section is to demonstrate the capabilities of the phase-field approach to fracture within a contact situation. In particular, we investigate a quasi-static problem as well as a transient impact problem including complex three dimensional fracture patterns.

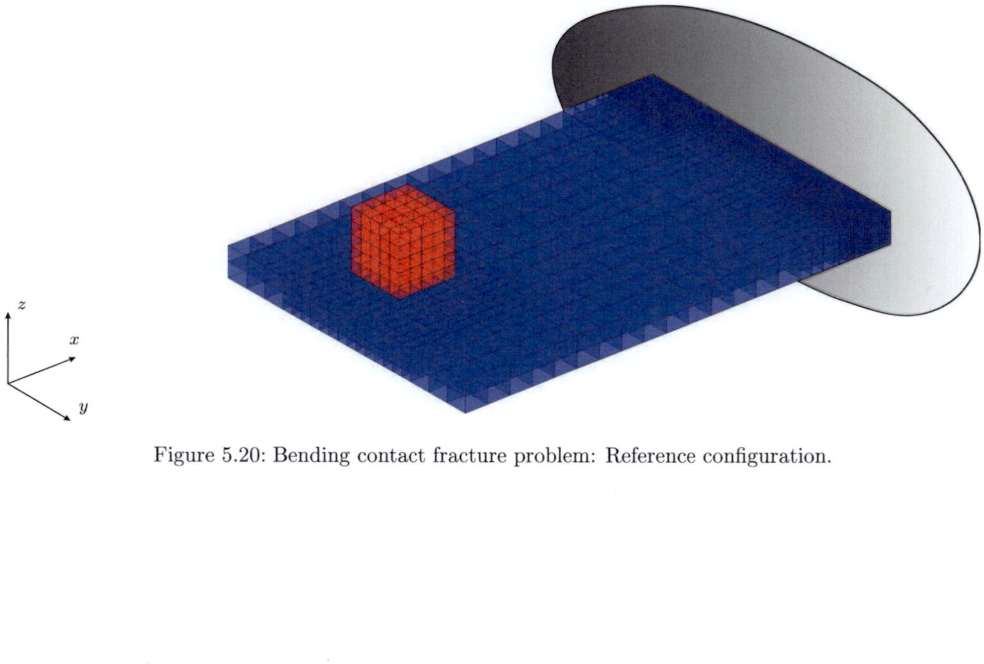

Figure 5.20: Bending contact fracture problem: Reference configuration.

Bending fracture problem

In this example, we consider a deformable block to be in contact with an elastic plate, see Figure 5.20 for the initial configuration. The plate is clamped and locally refined using a three level hierarchical refinement on the right hand side, since we expect peak stresses within this area. Moreover, the contact area is locally refined as well.

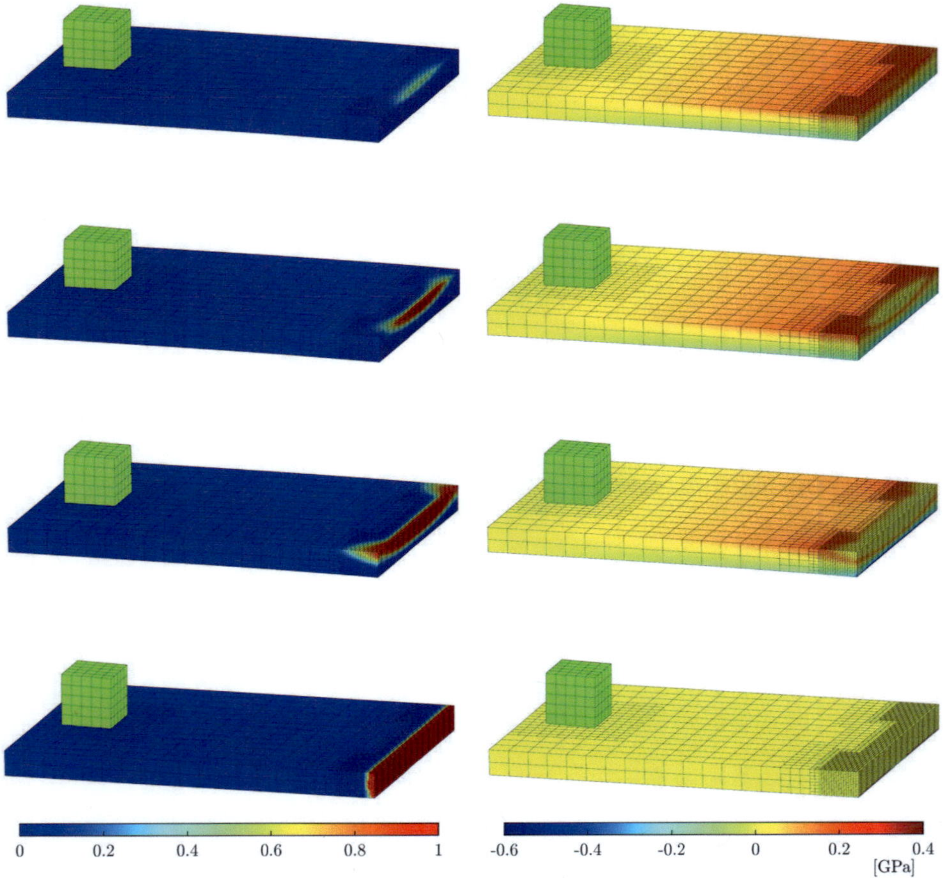

Figure 5.21: Bending contact fracture problem: Phase-field (left) and σ_{xx} stress distribution (right) after 140, 150, 155 and 160 quasi-static time steps (from top to bottom).

The plate is of size 0.3m$\times$0.2m$\times$0.02m, whereas the block is of size 0.04m$\times$0.04m$\times$0.04m. The center point of the block is placed 0.265m away form the clamping in negative x-direction. For both bodies, we assume that the constitutive behavior is governed by the Mooney-Rivlin material law defined in (4.21). The material parameters of the plate correspond to an aluminum-like material and take the values $\alpha = 8.707$GPa, $\beta = 4.353$GPa and $\kappa = 72.917\text{GPa}$. The parameters of the block are given by $\alpha = 0.011494\text{GPa}$, $\beta = 0.005747\text{GPa}$ and $\kappa = 0.333\text{GPa}$, which correspond to a synthetic substance with Young's modulus of $E = 0.1\text{GPa}$ and a Poisson ratio of $\nu = 0.45$. In addition, the phase-field parameters for the plate are chosen as $g_c = 2400\text{J/m}^2$, $l = 1.974 \times 10^{-3}\text{m}$ and $a_g = 0.1$.

The upper surface of the block is moved downwards (negative z-direction) with a constant increment size of $\Delta u_z = -1 \times 10^{-3}\text{m}$ for the first 10 quasi-static time steps until the contact is established, subsequently the increment size is reduced to $\Delta u_z = -0.05 \times 10^{-3}\text{m}$ for the next 126 time steps and finally, after the crack is initialized, the increment size is further reduced to $\Delta u_z = -0.01 \times 10^{-3}\text{m}$ for the remaining steps.

The block consists of $5 \times 5 \times 5$ quadratic B-spline elements and the plate of $19 \times 13 \times 2$ quadratic B-spline elements on level zero. A local refinement is applied to the expected contact area on the plate using a one level refinement. Furthermore, a three level refinement is applied to the Dirichlet boundary. In total 22261 elements with overall 107864 degrees of freedom are used. Note that a globally refined plate would require more than a million degrees of freedom without improvements on the accuracy of the areas of interest.

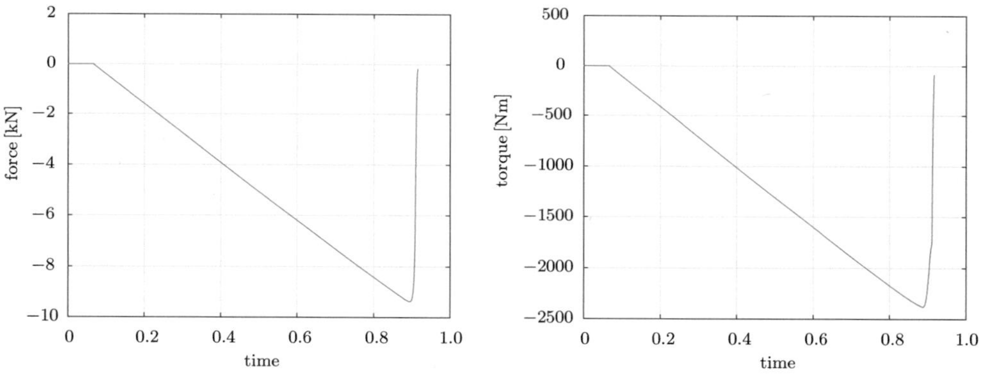

Figure 5.22: Bending contact fracture problem: Applied load on the upper boundary of the block (left) and resulting torque at the clamping of the plate (right).

The phase-field as well as the σ_{xx} stress distribution is depicted in Figure 5.21 for different time steps. As can be observed, the plate is nearly completely ripped out of the bearing. We stopped the simulation at this point, since the plate simply becomes statically undetermined. The applied load on the upper surface of the block as well as the resulting torque on the clamping of the plate is plotted over time in Figure 5.22.

Impact fracture problem

In this final example we consider a transient impact problem in conjunction with the proposed phase-field approach to fracture, see Figure 5.23 for the reference configuration of the bodies in contact. The initial velocity of the upper, wedge-shaped body is $\boldsymbol{v}_0 = [0,\ 0,\ -100]\text{m/s}$, whereas the long edges of the plate are fixed in space.

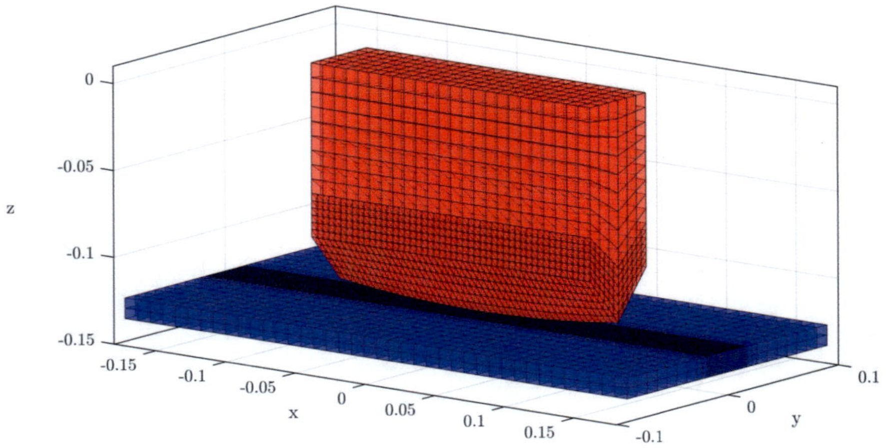

Figure 5.23: Impact fracture problem: Reference configuration.

The wedge-shaped body is approximately of dimension $0.2\text{m}\times 0.05\text{m}\times 0.134\text{m}$ and consists of $24\times 6\times 12$ quadratic B-spline elements on level zero. A one level hierarchical refinement is applied to the lower elements to refine the sharp-edged contact surface. The plate of size $0.36\text{m} \times 0.18\text{m} \times 0.018\text{m}$ consists of $40 \times 20 \times 2$ quadratic B-spline elements on level zero. A two level hierarchical refinement is used to achieve a sufficient mesh resolution of the possible impact region. The plate is equipped with the presented phase-field approach to fracture, thus we obtain in total 27632 elements with 129660 degrees of freedom.

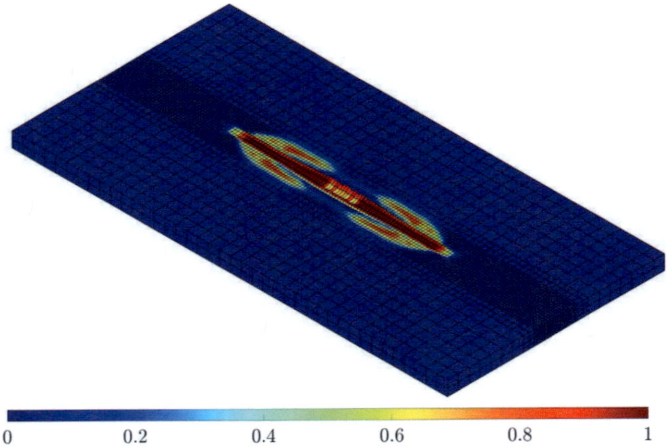

Figure 5.24: Impact fracture problem: Phase-field at time $t = 29 \times 10^{-6}\text{s}$.

The material behavior of both bodies is governed by a Neo-Hookean model, where the Lamé parameters of the wedge-shaped body correspond to steel-like material and take the values $\lambda = 121.15\text{GPa}$ and $\mu = 80.77\text{GPa}$. In addition the mass density is $\rho = 7850\text{kg/m}^3$. For the plate we apply acrylic glass with Lamé parameters $\lambda = 0.208\text{GPa}$ and $\mu = 0.073\text{GPa}$ and a mass density of $\rho = 1070\text{kg/m}^3$. Moreover, the phase-field parameters are set to $g_c = 350\text{J/m}^2$, $l = 0.0045\text{m}$ and $a_g = 2$.

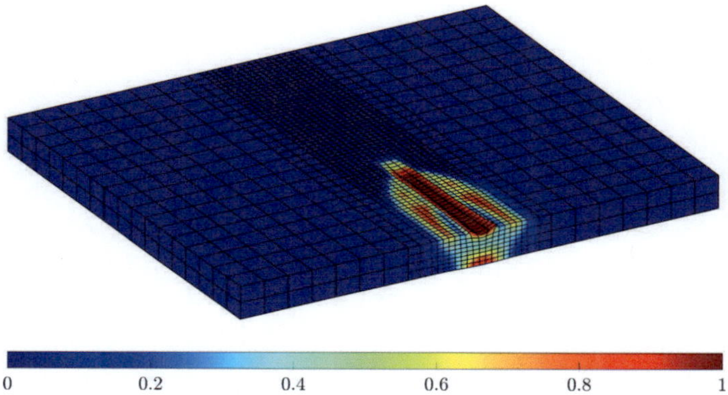

Figure 5.25: Impact fracture problem: Sectional view of plate at position -0.027m in longitudinal direction, phase-field at time $t = 29 \times 10^{-6}$s.

In Figure 5.24, the phase-field is shown for the plate after 145 constant time steps of size 0.2×10^{-6}s. A sectional view of the plate is given in Figure 5.25, whereas a detailed snapshot of the crack progression is shown in Figure 5.26. The latter figure displays the isosurface for $s = 0.7$ within the body. We can see that the surface in contact itself fractures with additional fracturing of the same surface parallel to the impact region. Moreover, due to bending of the lower surface additional cracks are initialized at the bottom of the plate as well and the crack fronts are merging within the plate.

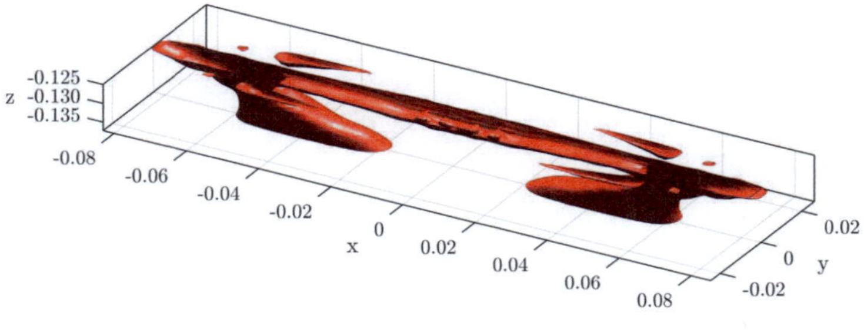

Figure 5.26: Impact fracture problem: Three dimensional crack propagation at time $t = 29 \times 10^{-6}$s, displaying the isosurface for $s = 0.7$.

This last example demonstrates the capabilities of the proposed phase-field methodology to deal with complex three dimensional crack patterns within an impact situation.

6 Summary and outlook

6.1 Summary

In the present work thermomechanical and fracture mechanical contact problems were examined using the concept of IGA. In particular, a hierarchical refinement scheme for higher order B-spline and NURBS based shape functions was proposed which allows to maintain the continuity of the unrefined spline bases as well as the partition of unity. The consistency as well as the convergence of the proposed refinement scheme have been shown by investigating a transient partial differential equation of fourth order.

For the thermomechanical system, a Hellinger-Reissner type variational principle was applied and compared with a displacement-temperature based approach. Extensive computational simulations have demonstrated the applicability and accuracy of the Hellinger-Reissner approach.

A novel polyconvex framework based on an anisotropic split of the principle invariants combined with a fourth order phase-field approach and a cubic degradation of the strain energy was considered for the fracture mechanical system. The superiority of the approach compared to previously established methods has been shown in a number of benchmark tests. Moreover, the capability to predict complex three dimensional crack patterns has also been demonstrated.

A novel mortar based contact formulation was considered for both, the thermomechanical system as well as the fracture mechanical system. To name a few innovations, a fast and reliable segmentation algorithm has been used for B-spline and NURBS discretized and hierarchical refined surfaces in contact and so-called triple mortar integrals have been introduced for the energy transfer between the mechanical and the thermal field. The capability of the entire framework could be shown in several numerical contact examples.

6.2 Outlook

Based on the present contribution the following research projects are currently in progress or seem to be worth investigating:

- A Hellinger-Reissner type variational principle for the proposed phase-field fracture system is currently in progress.

- First developments of a single multiphysical framework, combining thermomechanical systems with phase-field models for brittle and ductile fracture along with finite strain plasticity are also in progress.
- Enhanced time integration schemes for the Hellinger-Reissner approach are currently under investigation and need to be applied to the thermomechanical as well as to the fracture mechanical system.

A Tensor cross product

The cross product between two second order tensors $\boldsymbol{A}$ and $\boldsymbol{B}$ is defined as

$$[\boldsymbol{A} \times \boldsymbol{B}]_{IL} = \varepsilon_{IJK}\varepsilon_{LMN}A_{KL}B_{MN}, \tag{A.1}$$

where $\boldsymbol{\varepsilon}$ denotes the standard third order alternating tensor. The practical evaluation of this product reads

$$[\boldsymbol{A} \times \boldsymbol{B}] = \begin{bmatrix} [\boldsymbol{A} \times \boldsymbol{B}]_{11} & [\boldsymbol{A} \times \boldsymbol{B}]_{12} & [\boldsymbol{A} \times \boldsymbol{B}]_{13} \\ [\boldsymbol{A} \times \boldsymbol{B}]_{21} & [\boldsymbol{A} \times \boldsymbol{B}]_{22} & [\boldsymbol{A} \times \boldsymbol{B}]_{23} \\ [\boldsymbol{A} \times \boldsymbol{B}]_{31} & [\boldsymbol{A} \times \boldsymbol{B}]_{32} & [\boldsymbol{A} \times \boldsymbol{B}]_{33} \end{bmatrix} \tag{A.2}$$

with components

$$\begin{aligned}
[\boldsymbol{A} \times \boldsymbol{B}]_{11} &= A_{22}B_{33} - A_{23}B_{32} + A_{33}B_{22} - A_{32}B_{23} \\
[\boldsymbol{A} \times \boldsymbol{B}]_{12} &= A_{23}B_{31} - A_{21}B_{33} + A_{31}B_{23} - A_{33}B_{21} \\
[\boldsymbol{A} \times \boldsymbol{B}]_{13} &= A_{21}B_{32} - A_{22}B_{31} + A_{32}B_{21} - A_{31}B_{22} \\
[\boldsymbol{A} \times \boldsymbol{B}]_{21} &= A_{13}B_{32} - A_{12}B_{33} + A_{32}B_{13} - A_{33}B_{12} \\
[\boldsymbol{A} \times \boldsymbol{B}]_{22} &= A_{33}B_{11} - A_{31}B_{13} + A_{11}B_{33} - A_{13}B_{31} \\
[\boldsymbol{A} \times \boldsymbol{B}]_{23} &= A_{31}B_{12} - A_{32}B_{11} + A_{12}B_{31} - A_{11}B_{32} \\
[\boldsymbol{A} \times \boldsymbol{B}]_{31} &= A_{12}B_{23} - A_{13}B_{22} + A_{23}B_{12} - A_{22}B_{13} \\
[\boldsymbol{A} \times \boldsymbol{B}]_{32} &= A_{13}B_{21} - A_{11}B_{23} + A_{21}B_{13} - A_{23}B_{11} \\
[\boldsymbol{A} \times \boldsymbol{B}]_{33} &= A_{11}B_{22} - A_{12}B_{21} + A_{22}B_{11} - A_{21}B_{12}\,.
\end{aligned} \tag{A.3}$$

Bibliography

[1] C. Agelet de Saracibar. Numerical analysis of coupled thermomechanical frictional contact problems. Computational model and applications. *Archives of Computational Methods in Engineering*, 5(3):243–301, 1998.

[2] H. Altenbach. *Kontinuumsmechanik: Einführung in die materialunabhängigen und materialabhängigen Gleichungen.* Springer Berlin Heidelberg, 2012.

[3] D. Anders, M. Dittmann, and K. Weinberg. A higher-order finite element approach to the Kuramoto-Sivashinsky equation. *ZAMM - Journal of Applied Mathematics and Mechanics / Zeitschrift für Angewandte Mathematik und Mechanik*, 92(8):599–607, 2012.

[4] J.H. Argyris and J.S. Doltsinis. On the natural formulation and analysis of large deformation coupled thermomechanical problems. *Computer Methods in Applied Mechanics and Engineering*, 25(2):195–253, 1981.

[5] F. Armero and E. Petöcz. A new dissipative time-stepping algorithm for frictional contact problems: formulation and analysis. *Computer Methods in Applied Mechanics and Engineering*, 179(1–2):151–178, 1999.

[6] F. Armero and J.C. Simo. A new unconditionally stable fractional step method for non-linear coupled thermomechanical problems. *International Journal for Numerical Methods in Engineering*, 35(4):737–766, 1992.

[7] M. Bahrami, J.R. Culham, M.M. Yananovich, and G.E. Schneider. Review of thermal joint resistance models for nonconforming rough surfaces. *Applied Mechanics Reviews*, 59(1):1–12, 2006.

[8] J.M. Ball. Convexity conditions and existence theorems in nonlinear elasticity. *Archive for Rational Mechanics and Analysis*, 63(4):337–403, 1976.

[9] Y. Bazilevs, V.M. Calo, J.A. Cottrell, J.A. Evans, T.J.R. Hughes, S. Lipton, M.A. Scott, and T.W. Sederberg. Isogeometric analysis using T-splines. *Computer Methods in Applied Mechanics and Engineering*, 199(5–8):229–263, 2010. Computational Geometry and Analysis.

[10] J. Bonet, A.J. Gil, and R. Ortigosa. A computational framework for polyconvex large strain elasticity. *Computer Methods in Applied Mechanics and Engineering*, 283:1061–1094, 2015.

[11] M.J. Borden. *Isogeometric analysis of phase-field models for dynamic brittle and ductile fracture.* PhD thesis, 2012.

[12] M.J. Borden, T.J.R. Hughes, C.M. Landis, and C.V. Verhoosel. A higher-order phase-field model for brittle fracture: Formulation and analysis within the isogeometric analysis framework. *Computer Methods in Applied Mechanics and Engineering*, 273:100–118, 2014.

[13] P.B. Bornemann and F. Cirak. A subdivision-based implementation of the hierarchical b-spline finite element method. *Computer Methods in Applied Mechanics and Engineering*, 253:584–598, 2013.

[14] E. Brivadis, A. Buffa, B. Wohlmuth, and L. Wunderlich. Isogeometric mortar methods. *Computer Methods in Applied Mechanics and Engineering*, 284:292–319, 2015. Isogeometric Analysis Special Issue.

[15] J.W. Cahn and J.E. Hilliard. Free Energy of a Nonuniform System. I. Interfacial Free Energy. *The Journal of Chemical Physics*, 28(2):258–267, 1958.

[16] G.T. Camacho and M. Ortiz. Computational modelling of impact damage in brittle materials. *International Journal of Solids and Structures*, 33(20–22):2899–2938, 1996.

[17] R. Clausius. Ueber die bewegende Kraft der Wärme und die Gesetze, welche sich daraus für die Wärmelehre selbst ableiten lassen. *Annalen der Physik*, 155(3):368–397, 1850.

[18] J.A. Cottrell, T.J.R. Hughes, and Y. Bazilevs. *Isogeometric Analysis: Toward Integration of CAD and FEA*. Wiley, New York, 2009.

[19] J.A. Cottrell, T.J.R. Hughes, and A. Reali. Studies of refinement and continuity in isogeometric structural analysis. *Computer Methods in Applied Mechanics and Engineering*, 196(41–44):4160–4183, 2007.

[20] J.A. Cottrell, A. Reali, Y. Bazilevs, and T.J.R. Hughes. Isogeometric analysis of structural vibrations. *Computer Methods in Applied Mechanics and Engineering*, 195(41–43):5257–5296, 2006. John H. Argyris Memorial Issue. Part {II}.

[21] R. Courant. Variational methods for the solution of problems of equilibrium and vibrations. *Bull. Amer. Math. Soc.*, 49(1):1–23, 01 1943.

[22] L. Beirão da Veiga, D. Cho, L.F. Pavarino, and S. Scacchi. Overlapping Schwarz Methods for Isogeometric Analysis. *SIAM Journal on Numerical Analysis*, 50(3):1394–1416, 2012.

[23] R. de Boer. *Vektor- und Tensorrechnung für Ingenieure*. Springer-Verlag Berlin, Heidelberg, 1982.

[24] C. de Boor. *A Practical Guide to Splines*. Springer-Verlag New York, 1st edition, 1978.

[25] L. De Lorenzis., İ. Temizer, P. Wriggers, and G. Zavarise. A large deformation frictional contact formulation using NURBS-based isogeometric analysis. *International Journal for Numerical Methods in Engineering*, 87(13):1278–1300, 2011.

[26] L. De Lorenzis, P. Wriggers, and G. Zavarise. A mortar formulation for 3D large deformation contact using NURBS-based isogeometric analysis and the augmented Lagrangian method. *Computational Mechanics*, 49(1):1–20, 2012.

[27] M. Dittmann, M. Franke, İ. Temizer, and C. Hesch. Isogeometric Analysis and thermomechanical Mortar contact problems. *Computer Methods in Applied Mechanics and Engineering*, 274:192–212, 2014.

[28] S. Doll and K. Schweizerhof. On the Development of Volumetric Strain Energy Functions. *Journal of Applied Mechanics*, 67(1):17–21, 1999.

[29] E.J. Evans, M.A. Scott, X. Li, and D.C. Thomas. Hierarchical T-splines: Analysis-suitability, Bézier extraction, and application as an adaptive basis for isogeometric analysis. *Computer Methods in Applied Mechanics and Engineering*, 284:1–20, 2015. Isogeometric Analysis Special Issue.

[30] D.R. Forsey and R.H. Bartels. Hierarchical B-spline Refinement. *SIGGRAPH Comput. Graph.*, 22(4):205–212, June 1988.

[31] G.A. Francfort and J.-J. Marigo. Revisiting brittle fracture as an energy minimization problem. *Journal of the Mechanics and Physics of Solids*, 46(8):1319–1342, 1998.

[32] M. Franke, C. Hesch, and P. Betsch. An augmentation technique for large deformation frictional contact problems. *International Journal for Numerical Methods in Engineering*, 94(5):513–534, 2013.

[33] H. Gomez and J. París. Numerical simulation of asymptotic states of the damped Kuramoto-Sivashinsky equation. *Phys. Rev. E*, 83:046702, Apr 2011.

[34] A.A. Griffith. The Phenomena of Rupture and Flow in Solids. *Philosophical Transactions of the Royal Society of London A: Mathematical, Physical and Engineering Sciences*, 221(582-593):163–198, 1921.

[35] E. Gürses and C. Miehe. A computational framework of three-dimensional configurational-force-driven brittle crack propagation. *Computer Methods in Applied Mechanics and Engineering*, 198(15–16):1413–1428, 2009.

[36] S. Hartmann and P. Neff. Polyconvexity of generalized polynomial-type hyperelastic strain energy functions for near-incompressibility. *International Journal of Solids and Structures*, 40(11):2767–2791, 2003.

[37] C. Hesch and P. Betsch. Transient three-dimensional domain decomposition problems: Frame-indifferent mortar constraints and conserving integration. *International Journal for Numerical Methods in Engineering*, 82(3):329–358, 2010.

[38] C. Hesch and P. Betsch. Energy-momentum consistent algorithms for dynamic thermomechanical problems—Application to mortar domain decomposition problems. *International Journal for Numerical Methods in Engineering*, 86(11):1277–1302, 2011.

[39] C. Hesch and P. Betsch. Transient 3d contact problems—NTS method: mixed methods and conserving integration. *Computational Mechanics*, 48(4):437–449, 2011.

[40] C. Hesch and P. Betsch. Transient three-dimensional contact problems: mortar method. Mixed methods and conserving integration. *Computational Mechanics*, 48(4):461–475, 2011.

[41] C. Hesch and P. Betsch. Isogeometric analysis and domain decomposition methods. *Computer Methods in Applied Mechanics and Engineering*, 213–216:104–112, 2012.

[42] C. Hesch, M. Franke, M. Dittmann, and İ. Temizer. Hierarchical NURBS and a higher-order phase-field approach to fracture for finite-deformation contact problems. *Computer Methods in Applied Mechanics and Engineering*, 301:242–258, 2016.

[43] C. Hesch, A.J. Gil, R. Ortigosa, M. Dittmann, C. Bilgen, P. Betsch, M. Franke, A. Janz, and K. Weinberg. A framework for polyconvex large strain phase-field methods to fracture. *Computer Methods in Applied Mechanics and Engineering*, 2016. submitted.

[44] C. Hesch, S. Schuß, M. Dittmann, M. Franke, and K. Weinberg. Isogeometric analysis and hierarchical refinement for higher-order phase-field models. *Computer Methods in Applied Mechanics and Engineering*, 303:185–207, 2016.

[45] C. Hesch and K. Weinberg. Thermodynamically consistent algorithms for a finite-deformation phase-field approach to fracture. *International Journal for Numerical Methods in Engineering*, 99(12):906–924, 2014.

[46] G.A. Holzapfel. *Nonlinear Solid Mechanics*. John Wiley & Sons, 2001.

[47] G.A. Holzapfel and J.C. Simo. Entropy elasticity of isotropic rubber-like solids at finite strains. *Computer Methods in Applied Mechanics and Engineering*, 132(1):17–44, 1996.

[48] S. Hüeber and B.I. Wohlmuth. Thermo-mechanical contact problems on non-matching meshes. *Computer Methods in Applied Mechanics and Engineering*, 198(15–16):1338–1350, 2009.

[49] T.J.R. Hughes, J.A. Cottrell, and Y. Bazilevs. Isogeometric analysis: CAD, finite elements, NURBS, exact geometry and mesh refinement. *Computer Methods in Applied Mechanics and Engineering*, 194(39–41):4135–4195, 2005.

[50] K. Hutter. The foundations of thermodynamics, its basic postulates and implications. A review of modern thermodynamics. *Acta Mechanica*, 27(1):1–54, 1977.

[51] G.R. Irwin. Elasticity and plasticity: fracture. In S. Függe, editor, *Encyclopedia of Physics*, 1958.

[52] W. Jiang and J.E. Dolbow. Adaptive refinement of hierarchical B-spline finite elements with an efficient data transfer algorithm. *International Journal for Numerical Methods in Engineering*, 102(3-4):233–256, 2015.

[53] L. Johansson and A. Klarbring. Thermoelastic frictional contact problems: Modelling, finite element approximation and numerical realization. *Computer Methods in Applied Mechanics and Engineering*, 105(2):181–210, 1993.

[54] J.-Y. Kim and S.-K. Youn. Isogeometric contact analysis using mortar method. *International Journal for Numerical Methods in Engineering*, 89(12):1559–1581, 2012.

[55] R. Kraft. *Hierarchical B-splines.* Citeseer, 1994.

[56] T.A. Laursen. On the development of thermodynamically consistent algorithms for thermomechanical frictional contact. *Computer Methods in Applied Mechanics and Engineering*, 177(3–4):273–287, 1999.

[57] T.A. Laursen, M.A. Puso, and J. Sanders. Mortar contact formulations for deformable–deformable contact: Past contributions and new extensions for enriched and embedded interface formulations. *Computer Methods in Applied Mechanics and Engineering*, 205–208:3–15, 2012. Special Issue on Advances in Computational Methods in Contact Mechanicsdedicated to the memory of Professor J.A.C. Martins.

[58] J. Lu. Isogeometric contact analysis: Geometric basis and formulation for frictionless contact. *Computer Methods in Applied Mechanics and Engineering*, 200(5–8):726–741, 2011.

[59] C.V. Madhusudana. *Thermal Contact Conductance.* Springer, New York, 1996.

[60] L.E. Malvern. *Introduction to the mechanics of a continous medium.* Prentice Hall, 1969.

[61] J.E. Marsden and T.J.R. Hughes. *Mathematical Foundations of Elasticity.* Prentice-Hall, INC, 1983.

[62] M.E. Matzen, T. Cichosz, and M. Bischoff. A point to segment contact formulation for isogeometric, NURBS based finite elements. *Computer Methods in Applied Mechanics and Engineering*, 255:27–39, 2013.

[63] C. Miehe. Entropic thermoelasticity at finite strains. Aspects of the formulation and numerical implementation. *Computer Methods in Applied Mechanics and Engineering*, 120(3–4):243–269, 1995.

[64] C. Miehe, M. Hofacker, and F. Welschinger. A phase field model for rate-independent crack propagation: Robust algorithmic implementation based on operator splits. *Computer Methods in Applied Mechanics and Engineering*, 199(45–48):2765–2778, 2010.

[65] C. Miehe, L.-M. Schänzel, and H. Ulmer. Phase field modeling of fracture in multi-physics problems. Part I. Balance of crack surface and failure criteria for brittle crack propagation in thermo-elastic solids. *Computer Methods in Applied Mechanics and Engineering*, 294:449–485, 2015.

[66] C. Miehe, F. Welschinger, and M. Hofacker. Thermodynamically consistent phase-field models of fracture: Variational principles and multi-field FE implementations. *International Journal for Numerical Methods in Engineering*, 83(10):1273–1311, 2010.

[67] J.J. Moreau. *New Variational Techniques in Mathematical Physics*, chapter On Unilateral Constraints, Friction and Plasticity, pages 171–322. Springer Berlin Heidelberg, Berlin, Heidelberg, 2011.

[68] U. Nackenhorst. The ALE-formulation of bodies in rolling contact: Theoretical foundations and finite element approach. *Computer Methods in Applied Mechanics and Engineering*, 193(39–41):4299–4322, 2004. The Arbitrary Lagrangian-Eulerian Formulation.

[69] V.G. Oancea and T.A. Laursen. A finite element formulation of thermomechanical rate-dependent frictional sliding. *International Journal for Numerical Methods in Engineering*, 40(23):4275–4311, 1997.

[70] R. Ortigosa, A.J. Gil, J. Bonet, and C. Hesch. A computational framework for polyconvex large strain elasticity for geometrically exact beam theory. *Computational Mechanics*, 57(2):277–303, 2016.

[71] A. Pandolfi and M. Ortiz. An Efficient Adaptive Procedure for Three-Dimensional Fragmentation Simulations. *Engineering with Computers*, 18(2):148–159, 2002.

[72] A. Pandolfi and K. Weinberg. A numerical approach to the analysis of failure modes in anisotropic plates. *Engineering Fracture Mechanics*, 78(9):2052–2069, 2011.

[73] M. Paniconi and K.R. Elder. Stationary, dynamical, and chaotic states of the two-dimensional damped Kuramoto-Sivashinsky equation. *Phys. Rev. E*, 56:2713–2721, 1997.

[74] B.N.J. Persson, B. Lorenz, and A.I. Volokitin. Heat transfer between elastic solids with randomly rough surfaces. *The European Physical Journal E*, 31(1):3–24, 2010.

[75] L. Piegl and W. Tiller. *The NURBS Book*. Springer, 2nd edition, 2010.

[76] M.A. Puso and T.A. Laursen. A mortar segment-to-segment contact method for large deformation solid mechanics. *Computer Methods in Applied Mechanics and Engineering*, 193(6–8):601–629, 2004.

[77] M.A. Puso and T.A. Laursen. A mortar segment-to-segment frictional contact method for large deformations. *Computer Methods in Applied Mechanics and Engineering*, 193(45–47):4891–4913, 2004.

[78] M.A. Puso, T.A. Laursen, and J. Solberg. A segment-to-segment mortar contact method for quadratic elements and large deformations. *Computer Methods in Applied Mechanics and Engineering*, 197(6–8):555–566, 2008.

[79] S. Reese and S. Govindjee. Theoretical and Numerical Aspects in the Thermo-Viscoelastic Material Behaviour of Rubber-Like Polymers. *Mechanics of Time-Dependent Materials*, 1(4):357–396, 1997.

[80] D. Schillinger, L. Dedè, M.A. Scott, J.A. Evans, M.J. Borden, E. Rank, and T.J.R. Hughes. An isogeometric design-through-analysis methodology based on adaptive hierarchical refinement of NURBS, immersed boundary methods, and T-spline CAD surfaces. *Computer Methods in Applied Mechanics and Engineering*, 249–252:116–150, 2012. Higher Order Finite Element and Isogeometric Methods.

[81] J. Schröder and P. Neff. Invariant formulation of hyperelastic transverse isotropy based on polyconvex free energy functions. *International Journal of Solids and Structures*, 40(2):401–445, 2003.

[82] J. Schröder, P. Wriggers, and D. Balzani. A new mixed finite element based on different approximations of the minors of deformation tensors. *Computer Methods in Applied Mechanics and Engineering*, 200(49–52):3583–3600, 2011.

[83] M.A. Scott, X. Li, T.W. Sederberg, and T.J.R. Hughes. Local refinement of analysis-suitable T-splines. *Computer Methods in Applied Mechanics and Engineering*, 213–216:206–222, 2012.

[84] J.C. Simo. Numerical analysis and simulation of plasticity. In *Numerical Methods for Solids (Part 3) Numerical Methods for Fluids (Part 1)*, volume 6 of *Handbook of Numerical Analysis*, pages 183–499. Elsevier, 1998.

[85] J.C. Simo and C. Miehe. Associative coupled thermoplasticity at finite strains: Formulation, numerical analysis and implementation. *Computer Methods in Applied Mechanics and Engineering*, 98(1):41–104, 1992.

[86] L.F. Stazi, E. Budyn, J. Chessa, and T. Belytschko. An extended finite element method with higher-order elements for curved cracks. *Computational Mechanics*, 31(1):38–48, 2003.

[87] N. Strömberg, L. Johansson, and A. Klarbring. Derivation and analysis of a generalized standard model for contact, friction and wear. *International Journal of Solids and Structures*, 33(13):1817–1836, 1996.

[88] A. Suwannachit. *Thermomechanical analysis of tire rubber compounds in rolling contact.* PhD thesis, 2013.

[89] B.A. Szabò and I. Babuška. *Finite Element Analysis.* John Wiley and Sons, 1991.

[90] İ. Temizer. A mixed formulation of mortar-based frictionless contact. *Computer Methods in Applied Mechanics and Engineering*, 223–224:173–185, 2012.

[91] İ. Temizer and C. Hesch. Hierarchical NURBS in frictionless contact. *Computer Methods in Applied Mechanics and Engineering*, 299:161–186, 2016.

[92] İ. Temizer, P. Wriggers, and T.J.R. Hughes. Contact treatment in isogeometric analysis with NURBS. *Computer Methods in Applied Mechanics and Engineering*, 200(9–12):1100–1112, 2011.

[93] İ. Temizer, P. Wriggers, and T.J.R. Hughes. Three-dimensional mortar-based frictional contact treatment in isogeometric analysis with NURBS. *Computer Methods in Applied Mechanics and Engineering*, 209–212:115–128, 2012.

[94] M.J. Turner, R.W. Clough, H.C. Martin, and L.J. Topp. Stiffness and Deflection Analysis of Complex Structures. *Journal of the Aeronautical Sciences*, 23:805–823, 1956.

[95] A.-V. Vuong, C. Giannelli, B. Jüttler, and B. Simeon. A hierarchical approach to adaptive local refinement in isogeometric analysis. *Computer Methods in Applied Mechanics and Engineering*, 200(49–52):3554–3567, 2011.

[96] A.-V. Vuong, C. Heinrich, and B. Simeon. ISOGAT: A 2D tutorial MATLAB code for Isogeometric Analysis. *Computer Aided Geometric Design*, 27(8):644–655, 2010. Advances in Applied Geometry.

[97] K. Washizu. *Variational Methods in Elasticity & Plasticity*. Pergamon Press, 3nd edition, 1982.

[98] K. Weinberg and C. Hesch. A high-order finite deformation phase-field approach to fracture. *Continuum Mechanics and Thermodynamics*, pages 1–11, 2015.

[99] P. Wriggers. *Computational Contact Mechanics*. Springer-Verlag, 2nd edition, 2006.

[100] Y. Xu and C.-W. Shu. Local discontinuous Galerkin methods for the Kuramoto–Sivashinsky equations and the Ito-type coupled KdV equations. *Computer Methods in Applied Mechanics and Engineering*, 195(25–28):3430–3447, 2006. Discontinuous Galerkin Methods.

[101] B. Yang and T.A. Laursen. A contact searching algorithm including bounding volume trees applied to finite sliding mortar formulations. *Computational Mechanics*, 41(2):189–205, 2008.

[102] B. Yang and T.A. Laursen. A large deformation mortar formulation of self contact with finite sliding. *Computer Methods in Applied Mechanics and Engineering*, 197(6–8):756–772, 2008.

[103] G. Zavarise, P. Wriggers, E. Stein, and B.A. Schrefler. Real contact mechanisms and finite element formulation—a coupled thermomechanical approach. *International Journal for Numerical Methods in Engineering*, 35(4):767–785, 1992.

Schriftenreihe des Instituts für Mechanik
ISSN 2363-4936

Hrsg. von: Prof. Dr.-Ing. habil. Peter Betsch, Prof. Dr.-Ing. habil. Thomas Seelig

---

Band 1 Marlon Franke
Discretisation techniques for large deformation computational contact elastodynamics. 2014
ISBN 978-3-7315-0278-4

Band 2 Philipp Hempel
Constitutive modeling of amorphous thermoplastic polymers with special emphasis on manufacturing processes. 2016
ISBN 978-3-7315-0550-1

Band 3 Martin Helbig
Mehrskalenmodellierung von Schädigung in gummimodifizierten thermoplastischen Kunststoffen. 2017
ISBN 978-3-7315-0571-6

Band 4 Maik Dittmann
Isogeometric analysis and hierarchical refinement for multi-field contact problems. 2017
ISBN 978-3-7315-0616-4

Die Bände sind unter www.ksp.kit.edu als PDF frei verfügbar oder als Druckausgabe bestellbar.